2018 SQA Specimen and Past Papers with Answers

National 5
CHEMISTRY

National 5 CHEMISTRY

2017 & 2018 Exams
and 2017 Specimen Question Paper

HODDER
GIBSON
AN HACHETTE UK COMPANY

This book contains the official SQA 2017 and 2018 Exams, and the 2017 Specimen Question Paper for National 5 Chemistry, with associated SQA-approved answers modified from the official marking instructions that accompany the paper.

In addition the book contains study skills advice. This has been specially commissioned by Hodder Gibson, and has been written by experienced senior teachers and examiners in line with the new National 5 syllabus and assessment outlines. This is not SQA material but has been devised to provide further guidance for National 5 examinations.

Hodder Gibson is grateful to the copyright holders, as credited on the final page of the Answer section, for permission to use their material. Every effort has been made to trace the copyright holders and to obtain their permission for the use of copyright material. Hodder Gibson will be happy to receive information allowing us to rectify any error or omission in future editions.

Hachette UK's policy is to use papers that are natural, renewable and recyclable products and made from wood grown in sustainable forests. The logging and manufacturing processes are expected to conform to the environmental regulations of the country of origin.

Orders: please contact Bookpoint Ltd, 130 Park Drive, Milton Park, Abingdon, Oxon OX14 4SE. Telephone: (44) 01235 827827. Fax: (44) 01235 400454. Lines are open 9.00–5.00, Monday to Saturday, with a 24-hour message answering service. Visit our website at www.hoddereducation.co.uk. Hodder Gibson can also be contacted directly at hoddergibson@hodder.co.uk

This collection first published in 2018 by
Hodder Gibson, an imprint of Hodder Education,
An Hachette UK Company
211 St Vincent Street
Glasgow G2 5QY

Typeset by Aptara, Inc.

Printed in the UK

A catalogue record for this title is available from the British Library

ISBN: 978-1-5104-5661-7

2 1

2019 2018

MIX
Paper from
responsible sources
FSC™ C104740

Introduction

National 5 Chemistry

This book of SQA past papers contains the question papers used in the 2017 and 2018 exams (with answers at the back of the book). A specimen question paper reflecting the content and duration of the exam in 2018 is also included.

All of the question papers included in the book provide excellent representative exam practice for the final exams. Using these papers as part of your revision will help you to develop the vital skills and techniques needed for the exam, and will help you identify any knowledge gaps you may have.

It is always a very good idea to refer to SQA's website for the most up-to-date course specification documents. These are available at www.sqa.org.uk/sqa/47428

The course

National 5 courses have changed.

Unit Assessments have been removed from all National 5 courses. This means that you do not have to pass Unit Assessments in order to be eligible to take the exam. It also means that the exams for 2018 onwards have been updated and strengthened to assess your knowledge and skills across the whole course.

To achieve a pass in National 5 Chemistry, there are two main components you must complete.

Component 1 – Assignment

You are required to submit an assignment that is worth 20% (20 marks) of your final grade. This assignment will be based on research but it is important that you are aware that practical/experimental/field work is also a mandatory feature of your assignment so has to be included. This assignment requires you to apply skills, knowledge and understanding to investigate a relevant topic in chemistry and its effect on the environment and/or society. Your school or college will provide you with a Candidate's Guide for this assignment, which has been produced by the SQA. This guide gives information on what is required to complete the report and gain as many marks as possible. When you have completed your research and collected your experimental data, you will be allocated a maximum of 1 hour 30 minutes to complete your report in school.

Your assignment report will be marked by the SQA.

Component 2 – The question paper

The question paper will assess breadth and depth of knowledge and understanding from across the whole course. It is worth 80% of your final grade and is marked out of 100 possible marks. The question paper will require you to:

- make statements, provide explanations, and describe information to demonstrate knowledge and understanding.
- apply knowledge and understanding to new situations to solve problems.
- plan and design experiments.
- present information in various forms such as graphs, tables, etc.
- perform calculations based on information given.
- give predictions or make generalisations based on information given.
- draw conclusions based on information given.
- suggest improvement to experiments to improve the accuracy of results obtained or to improve safety.

To achieve a "C" grade in National 5 Chemistry you must achieve about 50% of the 120 marks available when the two components, i.e. the question paper and the assignment are combined. For a "B" you will need 60%, while for an "A" grade you must ensure that you gain as many of the available marks as possible, and at least 70%.

Each SQA Past Paper consists of two sections (a marking scheme for each section is provided at the end of this book):

- Section 1 will contain objective questions (multiple choice) and will have 25 marks.
- Section 2 will contain restricted and extended-response questions and will have 75 marks.

Each SQA Past Paper contains a variety of questions including some that require:

- demonstration and application of knowledge, and understanding of the mandatory content of the course from across the three areas of the course: chemical changes and structure; nature's chemistry; and chemistry in society.
- application of scientific inquiry skills.

How to use this book

This book can be used in two ways:

1. You can complete an entire paper under exam conditions, without the use of books or notes, and then mark the papers using the marking scheme provided. This method gives you a clear indication of the level you are working at and should highlight the content areas that you need to work on before attempting the next paper. This method also allows you to see your progress as you complete each paper.

2. You can complete a paper using your notes and books. Try the question first and then refer to your notes if you are unable to answer the question. This is a form of studying and by doing this you will cover all the areas of content that you are weakest in. You should notice that you are referring to your notes less with each paper completed.

Try to practise as many questions as possible. This will get you used to the language used in the papers and ultimately improve your chances of success.

Some hints and tips

Below is a list of hints and tips that will help you to achieve your full potential in the National 5 exam.

- Ensure that you **read each question carefully**. Scanning the question and missing the main points results in mistakes being made. Some students highlight the main points of a question with a highlighter pen to ensure that they don't miss anything out.

- Open ended questions include the statement **"Using your knowledge of chemistry"**. These questions provide you with an opportunity to show off your chemistry knowledge. To obtain the three marks on offer for these questions, you must demonstrate a good understanding of the chemistry involved and provide a logically correct answer to the question posed.

- When doing calculations, ensure that you **show all of your working**. If you make a simple arithmetical mistake you may still be awarded some of the marks, but only if your working is laid out clearly so that the examiner can see where you went wrong and what you did correctly. Just giving the answers is very risky so you should always show your working.

- **Attempt all questions.** Giving no answer at all means that you will definitely not gain any marks.

- When you are required to read a passage to answer a question, ensure that you **read it carefully** as the information you require is contained within it. It may not be obvious at first, but the answers will be contained within the passage.

- If you are asked to "explain" in a question, then you must **explain your answer fully**. For example, if you are asked to explain how a covalent bond holds atoms together then you cannot simply say:

    "A covalent bond is a shared pair of electrons between atoms in a non-metal."

This answer tells the examiner what a covalent bond is, but does not explain how it holds the atoms together. To gain the marks, an answer similar to this should be written:

    "A covalent bond is a shared pair of electrons between atoms in a non-metal. The shared electrons are attracted to the nuclei of both atoms, which creates a tug-of-war effect, creating the covalent bond."

- You may be required to draw one graph in each exam. To obtain all the marks, ensure that the graphs have **labels**, **units**, **points plotted correctly** and a line of "best fit" drawn between the points.

- Use your **data booklet** when you are asked to write formulas, ionic formulas, formula mass, etc. You have the data booklet in front of you so use it to double check the numbers you require.

- Work on your **timing**. The multiple-choice section (Section 1) should take approximately 40 minutes. Attempt to answer the multiple-choice questions before you look at the four possible answers, as this will improve your confidence. Use scrap paper when required to scribble down structural formulae, calculations, chemical formulae, etc., as this will reduce your chances of making errors. If you are finding the question difficult, try to eliminate the obviously wrong answers to increase your chances.

- When asked to **predict or estimate** based on information from a graph or a table, then take your time to look for patterns. For example, if asked to predict a boiling point, try to establish if there is a regular change in boiling point and use that regular pattern to establish the unknown boiling point.

- When drawing a **diagram** of an experiment ask yourself the question, "Would this work if I set it up exactly like this in the lab?" Ensure that the method you have drawn would produce the desired results *safely*. If, for example, you are heating a flammable reactant such as alcohol then you will not gain the marks if you heat it with a Bunsen burner in your diagram; a water bath would be much safer! Make sure your diagram is labelled clearly.

Good luck!

Remember that the rewards for passing National 5 Chemistry are well worth it! Your pass will help you get the future you want for yourself. In the exam, be confident in your own ability. If you're not sure how to answer a question, trust your instincts and just give it a go anyway. Keep calm and don't panic! GOOD LUCK!

Study Skills – what you need to know to pass exams!

General exam revision: 20 top tips

When preparing for exams, it is easy to feel unsure of where to start or how to revise. This guide to general exam revision provides a good starting place, and, as these are very general tips, they can be applied to all your exams.

1. Start revising in good time.

Don't leave revision until the last minute – this will make you panic and it will be difficult to learn. Make a revision timetable that counts down the weeks to go.

2. Work to a study plan.

Set up sessions of work spread through the weeks ahead. Make sure each session has a focus and a clear purpose. What will you study, when and why? Be realistic about what you can achieve in each session, and don't be afraid to adjust your plans as needed.

3. Make sure you know exactly when your exams are.

Get your exam dates from the SQA website and use the timetable builder tool to create your own exam schedule. You will also get a personalised timetable from your school, but this might not be until close to the exam period.

4. Make sure that you know the topics that make up each course.

Studying is easier if material is in manageable chunks – why not use the SQA topic headings or create your own from your class notes? Ask your teacher for help on this if you are not sure.

5. Break the chunks up into even smaller bits.

The small chunks should be easier to cope with. Remember that they fit together to make larger ideas. Even the process of chunking down will help!

6. Ask yourself these key questions for each course:

- Are all topics compulsory or are there choices?
- Which topics seem to come up time and time again?
- Which topics are your strongest and which are your weakest?

Use your answers to these questions to work out how much time you will need to spend revising each topic.

7. Make sure you know what to expect in the exam.

The subject-specific introduction to this book will help with this. Make sure you can answer these questions:

- How is the paper structured?
- How much time is there for each part of the exam?
- What types of question are involved? These will vary depending on the subject so read the subject-specific section carefully.

8. Past papers are a vital revision tool!

Use past papers to support your revision wherever possible. This book contains the answers and mark schemes too – refer to these carefully when checking your work. Using the mark scheme is useful; even if you don't manage to get all the marks available first time when you first practise, it helps you identify how to extend and develop your answers to get more marks next time – and of course, in the real exam.

9. Use study methods that work well for you.

People study and learn in different ways. Reading and looking at diagrams suits some students. Others prefer to listen and hear material – what about reading out loud or getting a friend or family member to do this for you? You could also record and play back material.

10. There are three tried and tested ways to make material stick in your long-term memory:

- Practising – e.g. rehearsal, repeating
- Organising – e.g. making drawings, lists, diagrams, tables, memory aids
- Elaborating – e.g. incorporating the material into a story or an imagined journey

11. Learn actively.

Most people prefer to learn actively – for example, making notes, highlighting, redrawing and redrafting, making up memory aids, or writing past paper answers. A good way to stay engaged and inspired is to mix and match these methods – find the combination that best suits you. This is likely to vary depending on the topic or subject.

12. Be an expert.

Be sure to have a few areas in which you feel you are an expert. This often works because at least some of them will come up, which can boost confidence.

13. Try some visual methods.

Use symbols, diagrams, charts, flashcards, post-it notes etc. Don't forget – the brain takes in chunked images more easily than loads of text.

14. Remember – practice makes perfect.

Work on difficult areas again and again. Look and read – then test yourself. You cannot do this too much.

15. Try past papers against the clock.

Practise writing answers in a set time. This is a good habit from the start but is especially important when you get closer to exam time.

16. Collaborate with friends.

Test each other and talk about the material – this can really help. Two brains are better than one! It is amazing how talking about a problem can help you solve it.

17. Know your weaknesses.

Ask your teacher for help to identify what you don't know. Try to do this as early as possible. If you are having trouble, it is probably with a difficult topic, so your teacher will already be aware of this – most students will find it tough.

18. Have your materials organised and ready.

Know what is needed for each exam:

- Do you need a calculator or a ruler?
- Should you have pencils as well as pens?
- Will you need water or paper tissues?

19. Make full use of school resources.

Find out what support is on offer:

- Are there study classes available?
- When is the library open?
- When is the best time to ask for extra help?
- Can you borrow textbooks, study guides, past papers, etc.?
- Is school open for Easter revision?

20. Keep fit and healthy!

Try to stick to a routine as much as possible, including with sleep. If you are tired, sluggish or dehydrated, it is difficult to see how concentration is even possible. Combine study with relaxation, drink plenty of water, eat sensibly, and get fresh air and exercise – all these things will help more than you could imagine. Good luck!

NATIONAL 5

2017

X713/75/02

**Chemistry
Section 1—Questions**

MONDAY, 8 MAY

1:00 PM – 3:00 PM

Instructions for the completion of Section 1 are given on *Page two* of your question and answer booklet X713/75/01.

Record your answers on the answer grid on *Page three* of your question and answer booklet.

You may refer to the Chemistry Data Booklet for National 5.

Before leaving the examination room you must give your question and answer booklet to the Invigilator; if you do not, you may lose all the marks for this paper.

SECTION 1

1. In a reaction, the mass lost in 30 seconds was 2 g.

    What is the average rate of reaction, in $g\,s^{-1}$, over this time?

    A $\dfrac{1}{30}$

    B $\dfrac{30}{2}$

    C $\dfrac{1}{2}$

    D $\dfrac{2}{30}$

2. An atom has 21 protons, 21 electrons and 24 neutrons.

    The atom has

    A   atomic number 24 and mass number 42

    B   atomic number 45 and mass number 21

    C   atomic number 21 and mass number 45

    D   atomic number 24 and mass number 45.

3. What is the charge on the zinc ion in zinc dichromate, $ZnCr_2O_7$?

    You may wish to use the data booklet to help you.

    A   2+

    B   2−

    C   1+

    D   1−

4. The table contains information about magnesium and magnesium chloride.

	Melting Point (°C)	Density (g cm⁻³)
Magnesium	650	1·74
Magnesium chloride	714	2·32

    When molten magnesium chloride is electrolysed at 730 °C the magnesium appears as a

    A   solid on the surface of the molten magnesium chloride

    B   solid at the bottom of the molten magnesium chloride

    C   liquid at the bottom of the molten magnesium chloride

    D   liquid on the surface of the molten magnesium chloride.

5. Which of the following compounds is a base?

    A   Sodium carbonate

    B   Sodium chloride

    C   Sodium nitrate

    D   Sodium sulfate

6. $AgNO_3(aq)$ + $KCl(aq)$ $\longrightarrow$ $AgCl(s)$ + $KNO_3(aq)$

    Which of the following are the spectator ions in this reaction?

    A Ag^+ and Cl^-

    B K^+ and NO_3^-

    C Ag^+ and NO_3^-

    D K^+ and Cl^-

7. $x\ H_2O_2$ $\longrightarrow$ $y\ H_2O$ + $z\ O_2$

    This equation will be balanced when

    A $x = 1$, $y = 2$ and $z = 2$

    B $x = 1$, $y = 1$ and $z = 2$

    C $x = 2$, $y = 2$ and $z = 1$

    D $x = 2$, $y = 2$ and $z = 2$.

8. 0·25 moles of a gas has a mass of 7 g.

    Which of the following could be the molecular formula for the gas?

    A C_2H_6

    B C_2H_4

    C C_3H_8

    D C_3H_6

9. Which of the following solutions contains the **least** number of moles of solute?

    A   100 cm^3 of 0·4 mol l^{-1} solution

    B   200 cm^3 of 0·3 mol l^{-1} solution

    C   300 cm^3 of 1·0 mol l^{-1} solution

    D   400 cm^3 of 0·5 mol l^{-1} solution

[Turn over

10. Which of the following could be the molecular formula for an alkane?

  A C_7H_{16}

  B C_7H_{14}

  C C_7H_{12}

  D C_7H_{10}

11. A student added bromine solution to compound **X** and compound **Y**.

Compound **X**   Compound **Y**

Which line in the table is correct?

| | Decolourises bromine solution immediately | |
	Compound **X**	Compound **Y**
A	no	no
B	no	yes
C	yes	yes
D	yes	no

12. A compound burns in air. The only products of the reaction are carbon dioxide, sulfur dioxide and water.

  The compound **must** contain

  A   carbon and sulfur only

  B   carbon and hydrogen only

  C   carbon, hydrogen and sulfur

  D   carbon, hydrogen, sulfur and oxygen.

13. Vinegar is a solution of

  A   ethanol

  B   methanol

  C   ethanoic acid

  D   methanoic acid.

Page four

14. A reaction is exothermic if

    A   energy is absorbed from the surroundings

    B   energy is released to the surroundings

    C   energy is required to start the reaction

    D   there is no energy change.

15. Which of the following diagrams could be used to represent the structure of copper?

    A

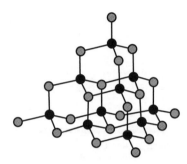

    B

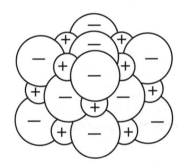

    C

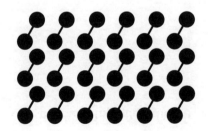

    D

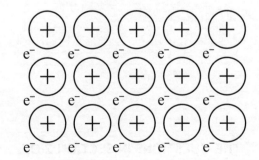

[Turn over

16. Which of the following metals is found uncombined in the Earth's crust?

    You may wish to use the data booklet to help you.

    A   Tin

    B   Magnesium

    C   Gold

    D   Sodium

17. Which of the following is **not** an essential element for healthy plant growth?

    A   Oxygen

    B   Nitrogen

    C   Potassium

    D   Phosphorus

18. The Haber process is the industrial process for the manufacture of

    A   nitric acid

    B   ammonia

    C   alkenes

    D   esters.

19. Which of the following salts can be prepared by a precipitation reaction?

    You may wish to use the data booklet to help you.

    A   Barium sulfate

    B   Lithium nitrate

    C   Calcium chloride

    D   Ammonium phosphate

20. A solution of accurately known concentration is more commonly known as a

    A   correct solution

    B   precise solution

    C   standard solution

    D   prepared solution.

[END OF SECTION 1. NOW ATTEMPT THE QUESTIONS IN SECTION 2 OF YOUR QUESTION AND ANSWER BOOKLET]

N5

National
Qualifications
2017

Mark

X713/75/01

Chemistry
Section 1—Answer Grid
And Section 2

MONDAY, 8 MAY

1:00 PM – 3:00 PM

Fill in these boxes and read what is printed below.

Full name of centre

Town

Forename(s)

Surname

Number of seat

Date of birth

Day   Month   Year   Scottish candidate number

Total marks — 80

SECTION 1 — 20 marks

Attempt ALL questions.

Instructions for the completion of Section 1 are given on *Page two*.

SECTION 2 — 60 marks

Attempt ALL questions.

You may refer to the Chemistry Data Booklet for National 5.

Write your answers clearly in the spaces provided in this booklet. Additional space for answers and rough work is provided at the end of this booklet. If you use this space you must clearly identify the question number you are attempting. Any rough work must be written in this booklet. You should score through your rough work when you have written your final copy.

Use **blue** or **black** ink.

Before leaving the examination room you must give this booklet to the Invigilator; if you do not, you may lose all the marks for this paper.

SECTION 1 — 20 marks

The questions for Section 1 are contained in the question paper X713/75/02.

Read these and record your answers on the answer grid on *Page three* opposite.

Use **blue** or **black** ink. Do NOT use gel pens or pencil.

1. The answer to each question is **either** A, B, C or D. Decide what your answer is, then fill in the appropriate bubble (see sample question below).

2. There is **only one correct** answer to each question.

3. Any rough working should be done on the additional space for answers and rough work at the end of this booklet.

Sample Question

To show that the ink in a ball-pen consists of a mixture of dyes, the method of separation would be

    A   fractional distillation

    B   chromatography

    C   fractional crystallisation

    D   filtration.

The correct answer is **B** — chromatography. The answer **B** bubble has been clearly filled in (see below).

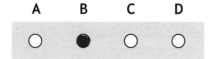

Changing an answer

If you decide to change your answer, cancel your first answer by putting a cross through it (see below) and fill in the answer you want. The answer below has been changed to **D**.

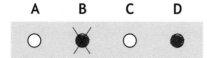

If you then decide to change back to an answer you have already scored out, put a tick (✓) to the **right** of the answer you want, as shown below:

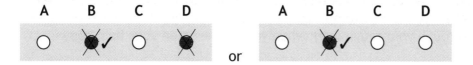

or

SECTION 1 — Answer Grid

	A	B	C	D
1	○	○	○	○
2	○	○	○	○
3	○	○	○	○
4	○	○	○	○
5	○	○	○	○
6	○	○	○	○
7	○	○	○	○
8	○	○	○	○
9	○	○	○	○
10	○	○	○	○
11	○	○	○	○
12	○	○	○	○
13	○	○	○	○
14	○	○	○	○
15	○	○	○	○
16	○	○	○	○
17	○	○	○	○
18	○	○	○	○
19	○	○	○	○
20	○	○	○	○

[BLANK PAGE]

DO NOT WRITE ON THIS PAGE

MARKS | DO NOT WRITE IN THIS MARGIN

SECTION 2 — 60 marks

Attempt ALL questions

1. A sample of argon contains three types of atom.

$$^{36}_{18}Ar \qquad ^{38}_{18}Ar \qquad ^{40}_{18}Ar$$

(a) State the term used to describe these different types of argon atom.

1

(b) Explain why the mass number of each type of atom is different.

1

(c) This sample of argon has an average atomic mass of 36·2.

State the mass number of the most common type of atom in the sample of argon.

1

[Turn over

MARKS DO NOT WRITE IN THIS MARGIN

2. Read the passage below and attempt the questions that follow.

Hydrogen Storage

The portable storage of hydrogen (H_2) is key to the development of hydrogen fuel cell cars. While many chemists focus their attention on the use of metal alloys and hydrides for storing hydrogen, others have investigated the potential use of carbon nanotubes.

A carbon nanotube is a tiny rolled up sheet of graphite. A research team has designed a pillared structure made up of vertical columns of carbon nanotubes which stabilise parallel graphene sheets. Graphene sheets are layers of carbon which are one atom thick.

Lithium atoms are added to the pillared structure to increase the hydrogen storage capacity. Researchers claim that one litre of the structure can store 41 g of hydrogen, which comes close to the US Department of Energy's target of 45 g.

Adapted from *InfoChem Magazine* (RSC), Nov 2008

(a) Name the term used to describe a tiny rolled up sheet of graphite.

1

(b) Name the metal added to the pillared structure to increase the hydrogen storage capacity.

1

(c) Calculate the number of moles of hydrogen that, researchers claim, can be stored by one litre of this structure.

2

Show your working clearly.

[Turn over for next question

DO NOT WRITE ON THIS PAGE

MARKS | DO NOT WRITE IN THIS MARGIN

3. Chlorine can form covalent and ionic bonds.

   (a) Chlorine gas is made up of diatomic molecules.

   Draw a diagram, showing all outer electrons, to represent a molecule of chlorine, Cl_2.

   1

   (b) Chloromethane is a covalent gas with a faint sweet odour.

   The structure of a chloromethane molecule is shown.

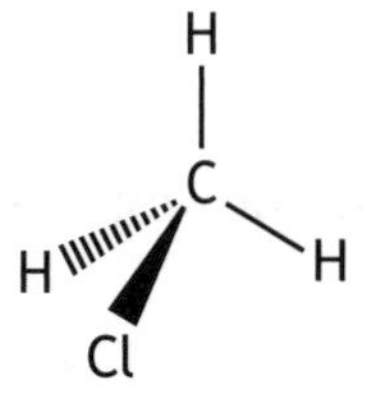

   State the name used to describe the shape of a molecule of chloromethane.

   1

MARKS | DO NOT WRITE IN THIS MARGIN

3.   (continued)

(c)   When chlorine reacts with sodium the ionic compound sodium chloride is formed.

A chloride ion has a stable electron arrangement.

Describe how a chlorine atom achieves this stable electron arrangement.

1

(d)   Covalent and ionic compounds have different physical properties.

Complete the table by circling the words which correctly describe the properties of the two compounds.

2

Compound	Melting point	Conductor of electricity
chloromethane gas	high/low	yes/no
solid sodium chloride	high/low	yes/no

[Turn over

MARKS | DO NOT WRITE IN THIS MARGIN

4. Iron is produced from iron ore in a blast furnace.

(a) Iron ore, limestone and carbon are added at the top of the blast furnace. Hot air is blown in near the bottom of the furnace and, through a series of chemical reactions, iron is produced. Waste gases are released near the top of the furnace. A layer of impurities is also produced which floats on top of the iron. The iron and impurities both flow off separately at the bottom of the furnace.

(i) Use this information to complete the diagram.   2

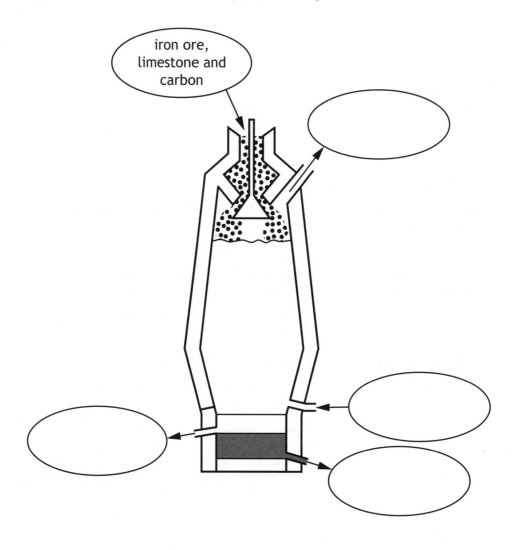

MARKS

DO NOT
WRITE IN
THIS
MARGIN

4. (a) (continued)

    (ii) Explain why the temperature at the bottom of the blast furnace should not drop below 1538 °C.   **1**

        You may wish to use the data booklet to help you.

  (b) Rusting occurs when iron is exposed to air and water.

    During rusting, iron initially loses two electrons to form iron(II) ions. These ions are further oxidised to form iron(III) ions.

    Write an ion-electron equation to show iron(II) ions forming iron(III) ions.   **1**

    You may wish to use the data booklet to help you.

[Turn over

MARKS | DO NOT WRITE IN THIS MARGIN

5. Phosphorus-32 is a radioisotope used in the detection of cancerous tumours.

(a) The graph shows how the percentage of phosphorus-32 in a sample changes over a period of time.

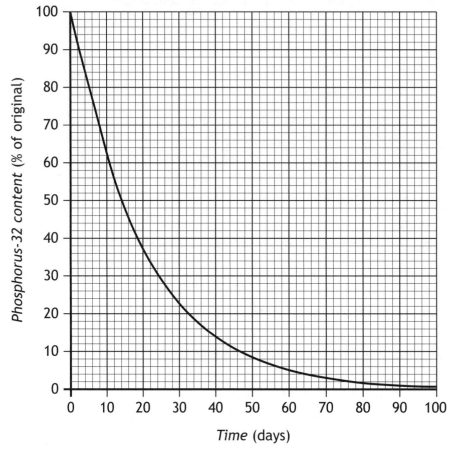

(i) Using the graph, calculate the half-life, in days, of phosphorus-32.   **1**

(ii) Using your answer to part (a) (i), calculate the time, in days, it would take for the mass of a 20 g sample of the radioisotope to decrease to 2·5 g.   **2**

(b) Phosphorus-32 decays by emitting radiation.

During this decay the atomic number increases by 1.

Name the type of radiation emitted when phosphorus-32 decays.   **1**

MARKS | DO NOT WRITE IN THIS MARGIN

6. A student wanted to investigate whether copper could be used as a catalyst for the reaction between zinc and sulfuric acid.

$$Zn(s) + H_2SO_4(aq) \longrightarrow ZnSO_4(aq) + H_2(g)$$

Using your knowledge of chemistry, suggest how the student could investigate this. 3

[Turn over

MARKS | DO NOT WRITE IN THIS MARGIN

7. Carboxylic acids can be used in household cleaning products.

(a) Name the functional group found in all carboxylic acids.

1

(b) Carboxylic acids have a range of physical and chemical properties. Melting point is an example of a physical property.

The table gives information about propanoic acid and butanoic acid.

Carboxylic acid	Melting point (°C)
propanoic acid	−21
butanoic acid	−5

(i) Draw a structural formula for butanoic acid.

1

(ii) Explain why butanoic acid has a higher melting point than propanoic acid.

2

MARKS | DO NOT WRITE IN THIS MARGIN

8. A teacher demonstrated the following experiment.

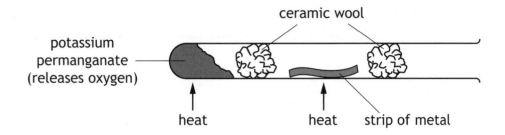

The results are shown in the table.

Metal	Observation
zinc	glowed brightly
copper	dull red glow
silver	no reaction

(a) (i) Describe what would be observed if the experiment was repeated using magnesium. **1**

(ii) The teacher repeated the experiment using copper powder.

State the effect this would have on the rate of the reaction between copper and oxygen. **1**

(b) Magnesium also reacts with steam to produce magnesium oxide and hydrogen gas.

$$Mg(s) \ + \ H_2O(g) \longrightarrow MgO(s) \ + \ H_2(g)$$

Identify the substance which is being oxidised. **1**

[**Turn over**

MARKS | DO NOT WRITE IN THIS MARGIN

9. The alkanes are a homologous series of saturated hydrocarbons.

(a) State what is meant by the term homologous series. **1**

(b) The structural formula of two alkanes is shown.

2-methylpentane   2,3-dimethylbutane

State the term used to describe a pair of alkanes such as 2-methylpentane and 2,3-dimethylbutane. **1**

[Turn over

MARKS | DO NOT WRITE IN THIS MARGIN

9. (continued)

(c) The alkanes present in a mixture were separated using a technique known as HPLC. The mixture was vaporised and then passed through a special column. Different alkanes take different amounts of time to pass through the column.

The results are shown.

Time taken to pass through the column

(i) Write a general statement linking the structure of the alkane to the length of time taken to pass through the column.   1

(ii) Propane was added to the mixture and the HPLC technique was repeated.

Draw an arrow on the graph to show the expected time taken for propane to pass through the column.   1

(An additional diagram, if required, can be found on *Page twenty-seven.*)

MARKS | DO NOT WRITE IN THIS MARGIN

10. A student set up an electrochemical cell using aluminium and copper electrodes as well as aluminium sulfate solution and copper(II) sulfate solution.

(a) (i) Complete the labels on the diagram to show the electrochemical cell which would give the direction of electron flow indicated. **1**

You may wish to use the data booklet to help you.

(An additional diagram, if required, can be found on *Page twenty-seven*.)

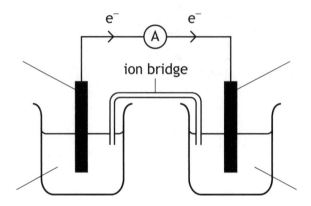

(ii) The two reactions which take place in the cell are

$$Al(s) \longrightarrow Al^{3+}(aq) + 3e^-$$

$$Cu^{2+}(aq) + 2e^- \longrightarrow Cu(s)$$

Write the redox equation for the overall reaction. **1**

(b) Calculate the percentage by mass of aluminium in aluminium sulfate, $Al_2(SO_4)_3$. **3**

Show your working clearly.

[Turn over

MARKS | DO NOT WRITE IN THIS MARGIN

11. Sulfur dioxide is an important industrial chemical.

Sulfur dioxide dissolves in water to produce sulfurous acid.

$$SO_2(g) \quad + \quad H_2O(\ell) \quad \longrightarrow \quad H_2SO_3(aq)$$

(a) Explain the change in the pH of the solution as sulfur dioxide dissolves.

2

(b) The graph shows the solubility of sulfur dioxide at different temperatures.

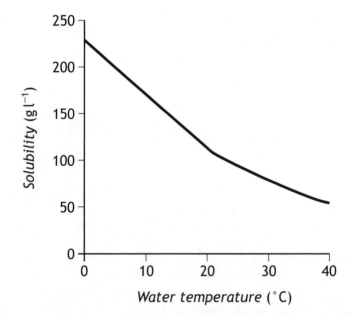

Describe the general trend in solubility as the temperature of the water increases.

1

[Turn over

MARKS | DO NOT WRITE IN THIS MARGIN

12. Geraniol is an essential oil known to have anti-inflammatory properties. A structure for the geraniol molecule is shown.

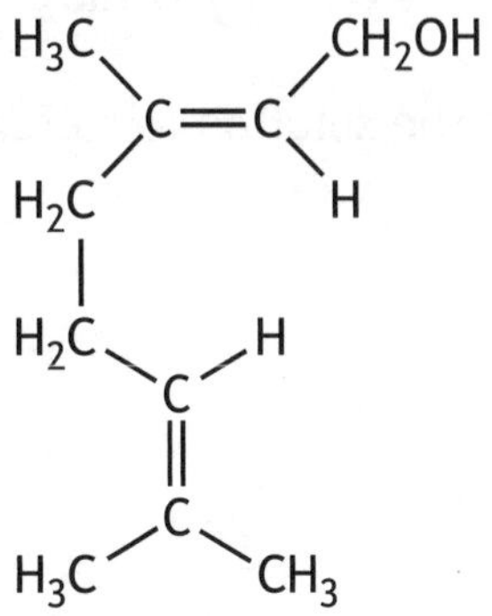

(a) Circle a functional group found in the geraniol molecule.   1

(An additional diagram, if required, can be found on *Page twenty-eight*.)

[Turn over

Page twenty

MARKS | DO NOT WRITE IN THIS MARGIN

12. (continued)

(b) One of the compounds used to flavour foods is geranyl propanoate.

Name the family to which geranyl propanoate belongs.   1

(c) A student prepared a sample of geranyl propanoate from geraniol and propanoic acid.

geraniol + propanoic acid $\longrightarrow$ geranyl propanoate + water

$C_{10}H_{18}O$ + $C_3H_6O_2$ $\longrightarrow$ $C_{13}H_{22}O_2$ + H_2O

15·4 g of geraniol was reacted with excess propanoic acid.

Calculate the mass, in grams, of geranyl propanoate which would be produced.   3

Show your working clearly.

[Turn over

MARKS | DO NOT WRITE IN THIS MARGIN

13. The alkynes are a family of hydrocarbons which contain a carbon to carbon triple bond. Three members of this family are shown.

propyne   but-1-yne   pent-1-yne

(a) Suggest a general formula for the alkyne family.   **1**

(b) Ethyne can undergo polymerisation to form poly(ethyne).

    (i) Draw the repeating unit in the polymer poly(ethyne).   **1**

    (ii) Name the type of polymerisation taking place when ethyne is converted to poly(ethyne).   **1**

[Turn over

MARKS | DO NOT WRITE IN THIS MARGIN

13.   (continued)

(c)   Alkynes can be prepared by reacting a dibromoalkane with potassium hydroxide solution.

$$H-\underset{\underset{H}{|}}{\overset{\overset{Br}{|}}{C}}-\underset{\underset{H}{|}}{\overset{\overset{Br}{|}}{C}}-\underset{\underset{H}{|}}{\overset{\overset{H}{|}}{C}}-H + 2KOH \longrightarrow H-C\equiv C-\underset{\underset{H}{|}}{\overset{\overset{H}{|}}{C}}-H + 2KBr + 2H_2O$$

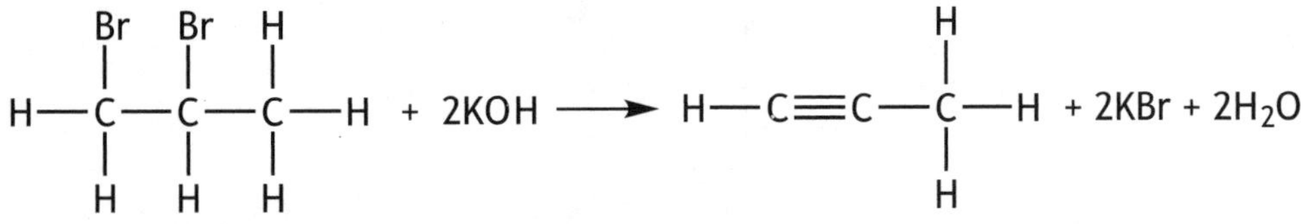

1,2-dibromopropane   propyne

(i)   Draw the **full** structural formula for the alkyne formed when 2,3-dibromobutane reacts with potassium hydroxide.   **1**

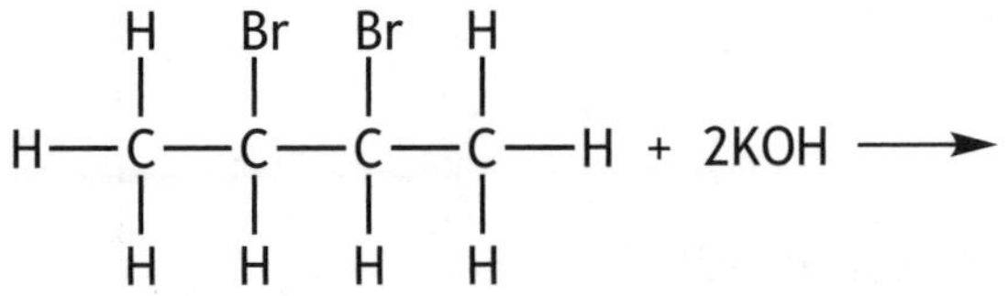

$$H-\underset{\underset{H}{|}}{\overset{\overset{H}{|}}{C}}-\underset{\underset{H}{|}}{\overset{\overset{Br}{|}}{C}}-\underset{\underset{H}{|}}{\overset{\overset{Br}{|}}{C}}-\underset{\underset{H}{|}}{\overset{\overset{H}{|}}{C}}-H + 2KOH \longrightarrow$$

2,3-dibromobutane

(ii)   The structure for 2,4-dibromopentane is shown below.

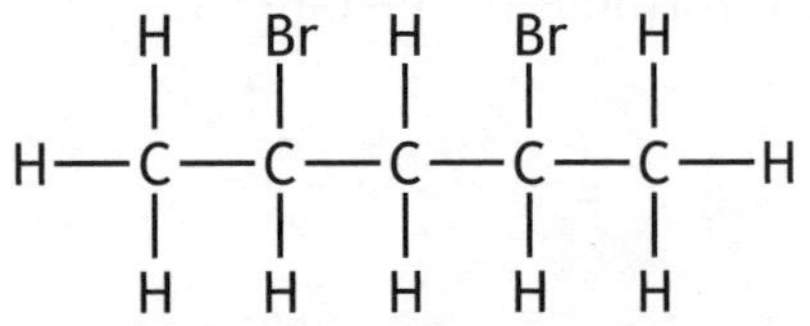

$$H-\underset{\underset{H}{|}}{\overset{\overset{H}{|}}{C}}-\underset{\underset{H}{|}}{\overset{\overset{Br}{|}}{C}}-\underset{\underset{H}{|}}{\overset{\overset{H}{|}}{C}}-\underset{\underset{H}{|}}{\overset{\overset{Br}{|}}{C}}-\underset{\underset{H}{|}}{\overset{\overset{H}{|}}{C}}-H$$

2,4-dibromopentane

Suggest a reason why 2,4-dibromopentane does **not** form an alkyne when it is added to potassium hydroxide solution.   **1**

[Turn over

MARKS | DO NOT WRITE IN THIS MARGIN

14. (a) A group of students carried out an experiment to measure the energy produced when 5 g samples of different alcohols were burned.

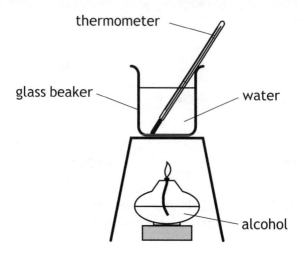

The results are shown.

Alcohol	Energy released (kJ)
propan-1-ol	158
butan-1-ol	170
pentan-1-ol	179
hexan-1-ol	185

(i) Draw a structural formula for hexan-1-ol.   1

(ii) Predict the energy released, in kJ, if the same mass of heptan-1-ol was burned.   1

[Turn over

MARKS DO NOT WRITE IN THIS MARGIN

14. (continued)

(b) The energy released when an alcohol burns can be used to heat liquids other than water.

The data below was collected when the energy released, by burning an alcohol, was used to heat a sodium chloride solution.

Energy released when the alcohol was burned (kJ)	13·3
Initial temperature of sodium chloride solution (°C)	15
Final temperature of sodium chloride solution (°C)	49
Mass of sodium chloride solution heated (g)	100

Calculate the specific heat capacity, in $kJ\,kg^{-1}\,°C^{-1}$, of the sodium chloride solution. **3**

You may wish to use the data booklet to help you.

Show your working clearly.

[Turn over for next question

MARKS | DO NOT WRITE IN THIS MARGIN

15. A student was given two solutions of sodium carbonate, one solution with a concentration of $0\cdot1\,mol\,l^{-1}$ and the other with a concentration of $0\cdot2\,mol\,l^{-1}$.

Using your knowledge of chemistry, suggest how the student could distinguish between the solutions.

3

[END OF QUESTION PAPER]

ADDITIONAL SPACE FOR ANSWERS

Additional diagram for Question 9 (c) (ii)

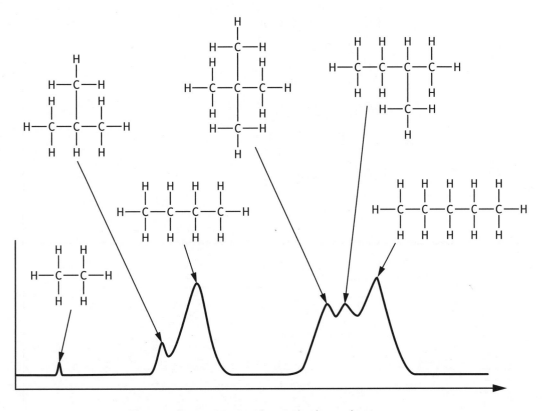

Time taken to pass through the column

Additional diagram for Question 10 (a) (i)

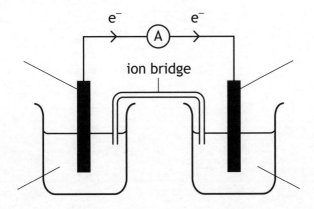

MARKS | DO NOT WRITE IN THIS MARGIN

ADDITIONAL SPACE FOR ANSWERS

Additional diagram for Question 12 (a)

MARKS | DO NOT WRITE IN THIS MARGIN

ADDITIONAL SPACE FOR ANSWERS AND ROUGH WORK

[BLANK PAGE]

DO NOT WRITE ON THIS PAGE

NATIONAL 5

2017 Specimen Question Paper

National
Qualifications
SPECIMEN ONLY

S813/75/02

**Chemistry
Section 1—Questions**

Date — Not applicable

Duration — 2 hours 30 minutes

Instructions for completion of Section 1 are given on *Page two* of your question and answer booklet S813/75/01.

Record your answers on the answer grid on *Page three* of your question and answer booklet.

You may refer to the Chemistry Data Booklet for National 5.

Before leaving the examination room you must give your question and answer booklet to the Invigilator; if you do not, you may lose all the marks for this paper.

SECTION 1 — 25 marks

Attempt ALL questions

1. Which of the following elements usually exists as diatomic molecules?

A   Helium

B   Nitrogen

C   Silicon

D   Sulfur

2. Which line in the table correctly describes a proton?

	Mass (atomic mass units)	Charge
A	negligible	+1
B	negligible	−1
C	1	+1
D	1	0

3. Ionic compounds conduct electricity when molten because they have

A   ions that are free to move

B   delocalised electrons

C   metal atoms

D   a lattice structure.

4. A molecule of phosphine is shown below.

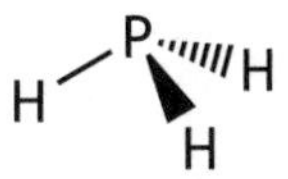

The shape of a molecule of phosphine is

A   linear

B   angular

C   tetrahedral

D   trigonal pyramidal.

5. The table gives information about some particles.

    Identify the particle which is a negative ion.

Particle	Number of		
	protons	neutrons	electrons
A	9	10	10
B	11	12	11
C	15	16	15
D	19	20	18

6. The table shows the colours of some ionic compounds in solution.

Compound	Colour
copper nitrate	blue
copper chromate	green
strontium nitrate	colourless
strontium chromate	yellow

The colour of the chromate ion is

A   blue

B   green

C   colourless

D   yellow.

7. Which of the following statements correctly describes the concentrations of $H^+(aq)$ and $OH^-(aq)$ ions in pure water?

A   The concentrations of $H^+(aq)$ and $OH^-(aq)$ ions are equal.

B   The concentrations of $H^+(aq)$ and $OH^-(aq)$ ions are zero.

C   The concentration of $H^+(aq)$ ions is greater than the concentration of $OH^-(aq)$ ions.

D   The concentration of $H^+(aq)$ ions is less than the concentration of $OH^-(aq)$ ions.

[Turn over

8.

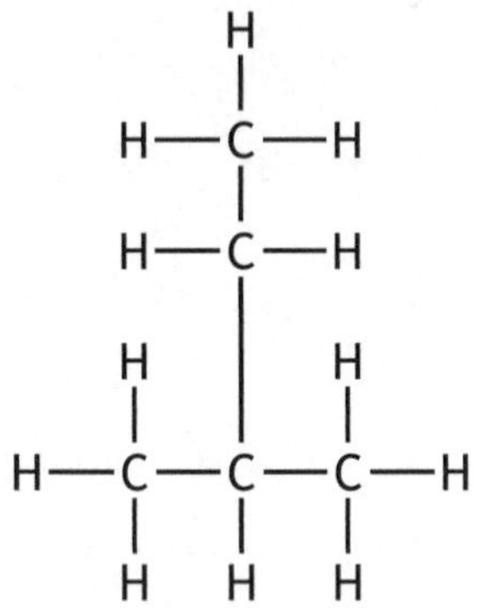

$$H-\underset{\underset{H}{|}}{\overset{\overset{H}{|}}{C}}-H$$

The name of the above compound is

A   2-ethylpropane

B   1,1-dimethylpropane

C   2-methylbutane

D   3-methylbutane.

9. Which of the following could be the molecular formula for a cycloalkane?

A C_6H_8

B C_6H_{10}

C C_6H_{12}

D C_6H_{14}

10. In which of the following types of reaction is oxygen a reactant?

A   Combustion

B   Neutralisation

C   Polymerisation

D   Precipitation

11. Molecules in which four different atoms are attached to a carbon atom are said to be chiral.

Which of the following molecules is chiral?

A

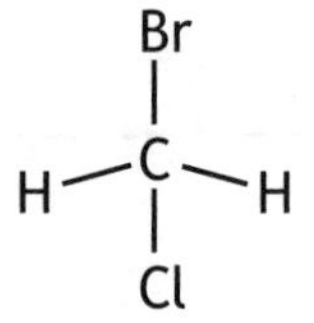

B

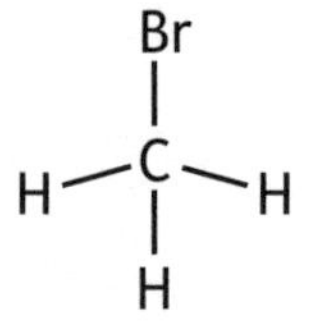

C

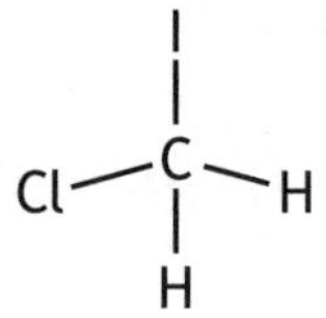

D

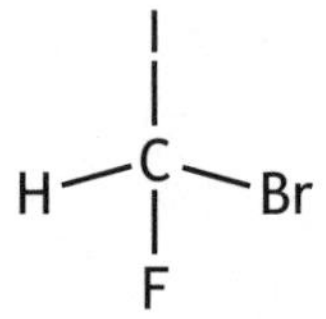

12. Three members of the cycloalkene family are

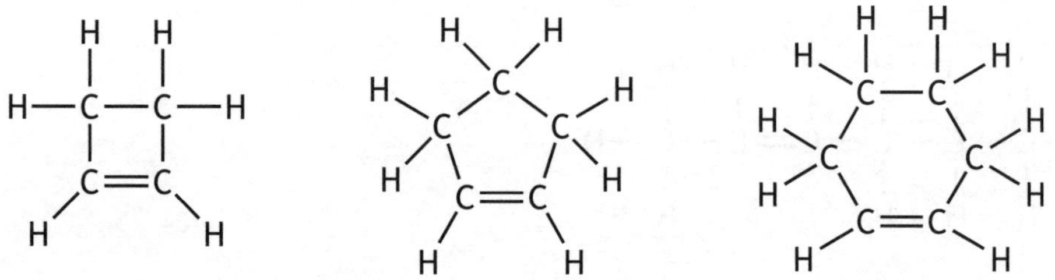

The general formula for the cycloalkene family is

A C_nH_{2n-2}

B C_nH_{2n-4}

C C_nH_{2n}

D C_nH_{2n+2}

[Turn over

13. Which of the following molecules is an isomer of hept-2-ene?

A

$$\begin{array}{ccccc} H & H & H & H & H \\ | & | & | & | & | \\ H-C-C-C-C-C-H \\ | & | & | & | & | \\ H & H & H & & H \\ & & & | & \\ & & & H-C-H \\ & & & | & \\ & & & H & \end{array}$$

B

$$\begin{array}{cccccc} H & H & H & H & H & H \\ | & | & | & | & | & | \\ H-C-C-C-C-C-C-H \\ | & | & | & | & | & | \\ H & H & H & & H & H \\ & & & | & \\ & & & H-C-H \\ & & & | & \\ & & & H & \end{array}$$

C

$$\begin{array}{c} H \\ | \\ H-C-H \\ | \\ \begin{array}{ccccc} H & H & H & H & H \\ | & | & | & | & | \\ H-C-C-C-C-C=C-H \\ | & | & | & | \\ H & H & H & H \end{array} \end{array}$$

D

$$\begin{array}{ccccccc} H & H & H & H & H & H & H \\ | & | & | & | & | & | & | \\ H-C-C-C-C-C=C-C-H \\ | & | & | & | & & | \\ H & H & H & H & & H \end{array}$$

14. A student tested some compounds. The results are given in the table.

Compound	pH of aqueous solution	Effect on bromine solution
	4	no effect
	4	decolourised
	7	no effect
	7	decolourised

Which line in the table below shows the correct results for the following compound?

	pH of aqueous solution	Effect on bromine solution
A	4	decolourised
B	7	decolourised
C	4	no effect
D	7	no effect

[Turn over

15. Which of the following diagrams could be used to represent the structure of a metal?

A

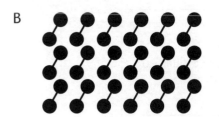

B

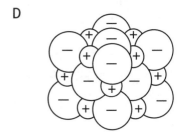

C

D

16. Which of the following substances does **not** produce water when it reacts with dilute acid?

A   Sodium hydroxide

B   Magnesium

C   Copper oxide

D   Ammonia solution

17. Which of the following metals can be extracted from its oxide by heat alone?

A   Aluminium

B   Zinc

C   Gold

D   Iron

18.

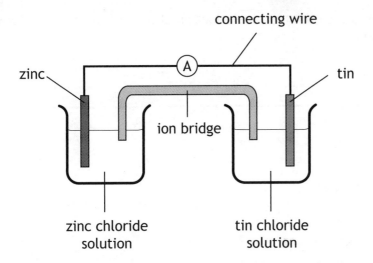

In the cell shown above, electrons flow through

A   the solution from tin to zinc

B   the solution from zinc to tin

C   the connecting wire from tin to zinc

D   the connecting wire from zinc to tin.

19.   Four cells were made by joining silver to copper, iron, tin and zinc.

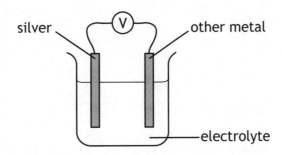

The voltages for the four cells are shown in the table.

Which cell contained silver joined to copper?

You may wish to use the data booklet to help you.

Cell	Voltage (V)
A	1·6
B	1·2
C	0·9
D	0·5

[Turn over

20. The ion-electron equation for the oxidation and reduction steps in the reaction between magnesium and silver(I) ions are:

$$Mg \rightarrow Mg^{2+} + 2e^-$$

$$Ag^+ + e^- \rightarrow Ag$$

The overall redox equation is

A $Mg + 2Ag^+ \rightarrow Mg^{2+} + 2Ag$

B $Mg + Ag^+ \rightarrow Mg^{2+} + Ag$

C $Mg + Ag^+ + e^- \rightarrow Mg^{2+} + Ag + 2e^-$

D $Mg + 2Ag \rightarrow Mg^{2+} + 2Ag^+$.

21. The structure below shows a section of an addition polymer.

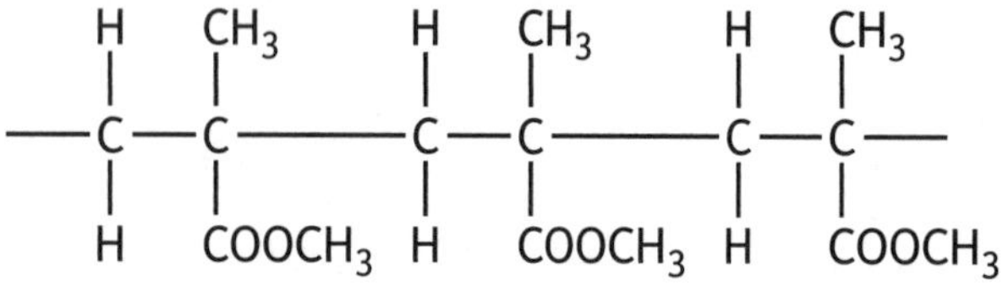

Which of the following molecules is used to make this polymer?

A

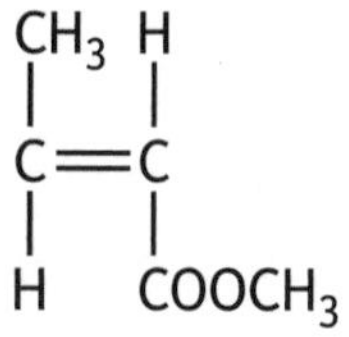

B

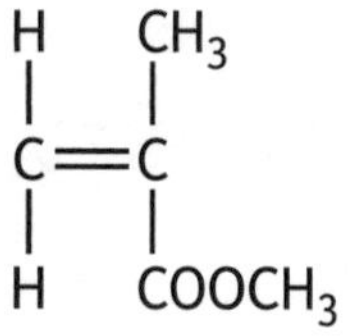

C

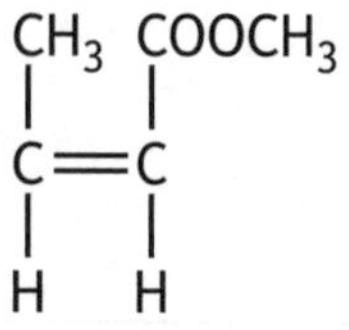

D

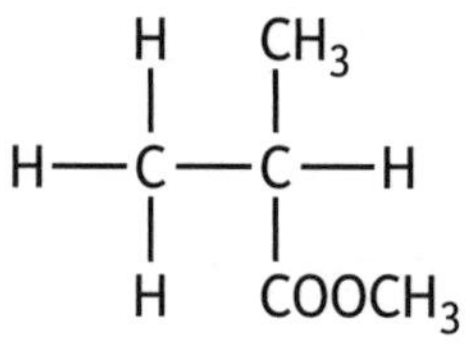

22. Hydrogen gas

    A   burns with a pop

    B   relights a glowing splint

    C   turns damp pH paper red

    D   turns limewater cloudy.

23. What is the charge on an iron ion in $Fe_2(SO_4)_3$?

    A   3–

    B   3+

    C   2–

    D   2+

24. Sodium sulfate solution reacts with barium chloride solution.

    $Na_2SO_4(aq)   +   BaCl_2(aq)   \rightarrow   BaSO_4(s)   +   2NaCl(aq)$

    The spectator ions present in this reaction are

    A Ba^{2+} and Cl^-

    B Ba^{2+} and SO_4^{2-}

    C Na^+ and Cl^-

    D Na^+ and SO_4^{2-}

[Turn over

25. But-1-ene is a colourless, insoluble gas which is more dense than air but less dense than water.

Which of the following diagrams shows the most appropriate apparatus for collecting and measuring the volume of but-1-ene?

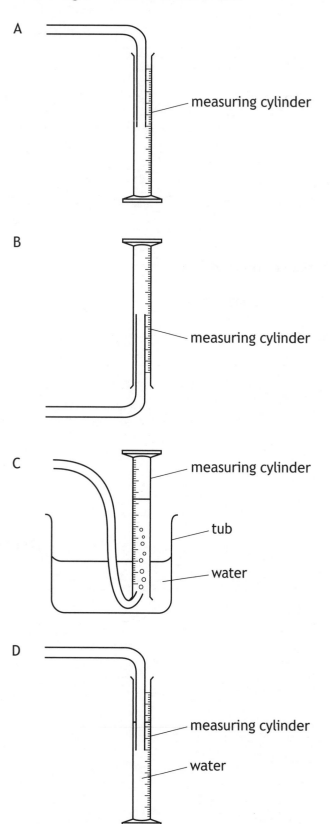

A — measuring cylinder

B — measuring cylinder

C — measuring cylinder
— tub
— water

D — measuring cylinder
— water

[END OF SECTION 1. NOW ATTEMPT THE QUESTIONS IN SECTION 2 OF YOUR QUESTION AND ANSWER BOOKLET]

N5

National Qualifications
SPECIMEN ONLY

Mark

S813/75/01

Chemistry
Section 1—Answer Grid
And Section 2

Date — Not applicable

Duration — 2 hours 30 minutes

Fill in these boxes and read what is printed below.

Full name of centre

Town

Forename(s)

Surname

Number of seat

Date of birth

Day	Month	Year

Scottish candidate number

Total marks — 100

SECTION 1 — 25 marks

Attempt ALL questions.

Instructions for the completion of Section 1 are given on *Page two*.

SECTION 2 — 75 marks

Attempt ALL questions.

You may refer to the Chemistry Data Booklet for National 5.

Write your answers clearly in the spaces provided in this booklet. Additional space for answers and rough work is provided at the end of this booklet. If you use this space you must clearly identify the question number you are attempting. Any rough work must be written in this booklet. Score through your rough work when you have written your final copy.

Use **blue** or **black** ink.

Before leaving the examination room you must give this booklet to the Invigilator; if you do not, you may lose all the marks for this paper.

SECTION 1 — 25 marks

The questions for Section 1 are contained in the question paper S813/75/02.

Read these and record your answers on the answer grid on *Page three* opposite.

Use **blue** or **black** ink. Do NOT use gel pens or pencil.

1. The answer to each question is **either** A, B, C, or D. Decide what your answer is, then fill in the appropriate bubble (see sample question below).

2. There is **only one correct** answer to each question.

3. Any rough working should be done on the additional space for answers and rough work at the end of this booklet.

Sample Question

To show that the ink in a ball-pen consists of a mixture of dyes, the method of separation would be

    A   fractional distillation

    B   chromatography

    C   fractional crystallisation

    D   filtration.

The correct answer is **B** — chromatography. The answer **B** bubble has been clearly filled in (see below).

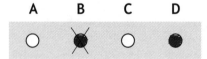

Changing an answer

If you decide to change your answer, cancel your first answer by putting a cross through it (see below) and fill in the answer you want. The answer below has been changed to **D**.

If you then decide to change back to an answer you have already scored out, put a tick (✓) to the **right** of the answer you want, as shown below:

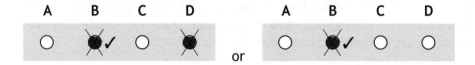

SECTION 1 — Answer Grid

	A	B	C	D
1	○	○	○	○
2	○	○	○	○
3	○	○	○	○
4	○	○	○	○
5	○	○	○	○
6	○	○	○	○
7	○	○	○	○
8	○	○	○	○
9	○	○	○	○
10	○	○	○	○
11	○	○	○	○
12	○	○	○	○
13	○	○	○	○
14	○	○	○	○
15	○	○	○	○
16	○	○	○	○
17	○	○	○	○
18	○	○	○	○
19	○	○	○	○
20	○	○	○	○
21	○	○	○	○
22	○	○	○	○
23	○	○	○	○
24	○	○	○	○
25	○	○	○	○

[BLANK PAGE]

DO NOT WRITE ON THIS PAGE

MARKS | DO NOT WRITE IN THIS MARGIN

SECTION 2 — 75 marks

Attempt ALL questions

1. Graphs can be used to show the change in the rate of a reaction as the reaction proceeds.

   The graph shows the volume of gas produced in an experiment over a period of time.

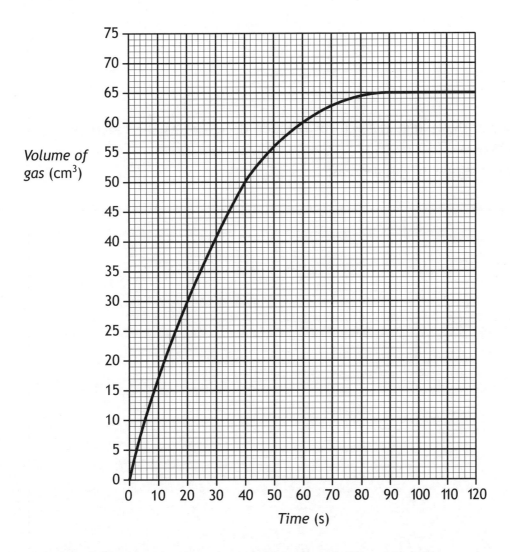

Volume of gas (cm^3)

Time (s)

   (a) State the time, in seconds, at which the reaction stopped.   1

[Turn over

MARKS | DO NOT WRITE IN THIS MARGIN

1.   (continued)

(b) Calculate the average rate of reaction for the first 20 seconds.   **3**
     Your answer must include the appropriate unit.
     Show your working clearly.

(c) The graph shows that the rate of reaction decreases as the reaction proceeds.

     Suggest a reason for this decrease.   **1**

MARKS | DO NOT WRITE IN THIS MARGIN

2. The group 7 element bromine was discovered by Balard in 1826.

Bromine gets its name from the Greek "bromos" meaning stench.

A sample of bromine consists of a mixture of two isotopes, $^{79}_{35}Br$ and $^{81}_{35}Br$.

(a) State what is meant by the term isotope.　　　1

(b) Complete the table for $^{79}_{35}Br$.　　　1

Isotope	Number of protons	Number of neutrons
$^{79}_{35}Br$		

(c) The sample of bromine has an average atomic mass of 80.

Suggest what this indicates about the amount of each isotope in this sample.　　　1

[Turn over

MARKS | DO NOT WRITE IN THIS MARGIN

2.   (continued)

(d) In 1825 bromine had been isolated from sea water by Liebig who mistakenly thought it was a compound of iodine and chlorine.

Using your knowledge of chemistry, comment on why Liebig might have made this mistake.

3

3. Antacid tablets are used to treat indigestion which is caused by excess acid in the stomach.

Different brands of tablets contain different active ingredients.

Name of active ingredient	magnesium carbonate	calcium carbonate	magnesium hydroxide	aluminium hydroxide
Reaction with acid	fizzes	fizzes	does not fizz	does not fizz
Cost per gram (pence)	16	11	7·5	22
Mass of solid needed to neutralise 20 cm^3 of acid (g)	0·7	1·2	0·6	0·4
Cost of neutralising 20 cm^3 of acid (pence)		13·2	4·5	8·8

(a) Write the formula, showing the charge on each ion, for aluminium hydroxide. **1**

(b) (i) Complete the table to show the cost of using magnesium carbonate to neutralise 20 cm^3 of acid. **1**

(ii) Using information from the table, state which **one** of the four active ingredients **you** would use to neutralise excess stomach acid.

Explain your choice. **1**

[Turn over

4. Sulfur dioxide gas is produced when fossil fuels containing sulfur are burned.

(a) When sulfur dioxide dissolves in water in the atmosphere "acid rain" is produced.

Circle the correct phrase to complete the sentence. **1**

Compared with pure water, acid rain contains $\left\{\begin{array}{l}\text{a higher} \\ \text{a lower} \\ \text{the same}\end{array}\right\}$ concentration of hydrogen ions.

(b) The table gives information about the solubility of sulfur dioxide.

Temperature (°C)	18	24	30	36	42	48
Solubility (g/100 cm³)	11·2	9·2	7·8	6·5	5·5	4·7

(i) Draw a graph of solubility against temperature.

Use appropriate scales to fill most of the graph paper. **4**

(Additional graph paper, if required, can be found on *Page twenty-eight*.)

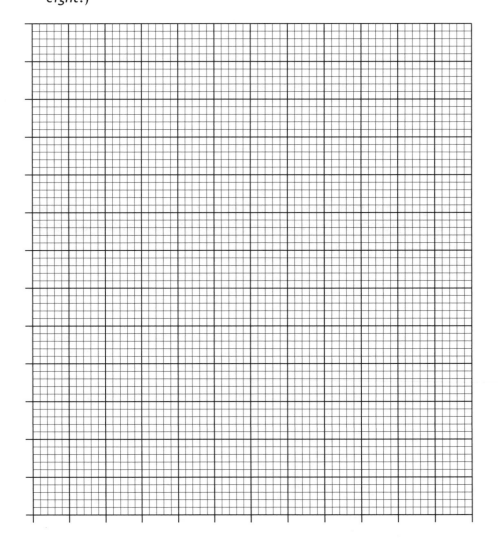

MARKS DO NOT WRITE IN THIS MARGIN

4. (b) (continued)

(ii) Estimate the solubility of sulfur dioxide, in g/100 cm^3, at 21 °C. 1

[Turn over

MARKS | DO NOT WRITE IN THIS MARGIN

5. A student investigated the reaction of carbonates with dilute hydrochloric acid.

    (a) In one reaction lithium carbonate reacted with dilute hydrochloric acid.
         The equation for the reaction is:

         $Li_2CO_3(s) + HCl(aq) \rightarrow LiCl(aq) + CO_2(g) + H_2O(\ell)$

         (i) Balance this equation.   1

         (ii) Identify the salt produced in this reaction.   1

    (b) In another reaction 1·0 g of calcium carbonate reacted with excess dilute hydrochloric acid.

         $CaCO_3(s) + 2HCl(aq) \rightarrow CaCl_2(aq) + CO_2(g) + H_2O(\ell)$

         (i) Calculate the mass, in grams, of carbon dioxide produced.   3
              Show your working clearly.

5.   (b)   (continued)

     (ii)   The student considered two methods to confirm the mass of carbon dioxide gas produced in this reaction.

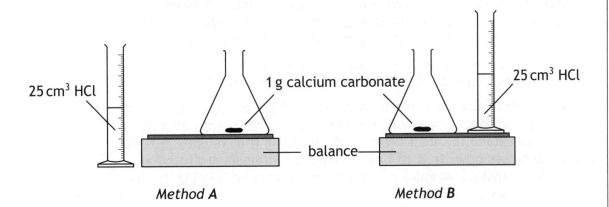

Method A   *Method B*

Method A	Method B
1. Add the acid from the measuring cylinder to the calcium carbonate in the flask.	1. Weigh the flask with the calcium carbonate and the acid in the measuring cylinder together.
2. Weigh the flask and contents.	2. Add the acid from the measuring cylinder to the calcium carbonate in the flask and replace the empty measuring cylinder on the balance.
3. Leave until no more bubbles are produced.	3. Leave until no more bubbles are produced.
4. Reweigh the flask and contents.	4. Reweigh the flask, contents and the empty measuring cylinder together.

Explain which method would give a more reliable estimate of the mass of carbon dioxide produced during the reaction.   2

[Turn over

MARKS | DO NOT WRITE IN THIS MARGIN

6. Read the passage below and answer the questions that follow.

Potassium Permanganate (KMnO₄)

Potassium permanganate's strong oxidising properties make it an effective disinfectant. Complaints such as athlete's foot and some fungal infections are treated by bathing the affected area in $KMnO_4$ solution.

In warm climates vegetables are washed in $KMnO_4$ to kill bacteria such as *E. coli*. Chemists use $KMnO_4$ in the manufacture of saccharin and benzoic acid.

Baeyer's reagent is an alkaline solution of $KMnO_4$ and is used to detect unsaturated organic compounds. The reaction of $KMnO_4$ with alkenes is also used to extend the shelf life of fruit. Ripening fruit releases ethene gas which causes other fruit to ripen. Shipping containers are fitted with gas scrubbers that use alumina or zeolite impregnated with $KMnO_4$ to stop the fruit ripening too quickly.

$$C_2H_4 + 4KMnO_4 \rightarrow 4MnO_2 + 4KOH + 2CO_2$$

Adapted from an article by Simon Cotton on "Soundbite molecules" in "Education in Chemistry" November 2009.

(a) Suggest an experimental test, including the result, to show that potassium is present in potassium permanganate.   1

You may wish to use the data booklet to help you.

(b) Suggest a pH for Baeyer's reagent.   1

(c) Name the gas removed by the scrubbers.   1

(d) Name a chemical mentioned in the passage which contains the following functional group.   1

(e) Zeolite is a substance that contains aluminium silicate.

Name the elements present in aluminium silicate.   1

MARKS | DO NOT WRITE IN THIS MARGIN

7.   In the 2012 London Olympics, alkanes were used as fuels for the Olympic flame.

(a)   The torches that carried the Olympic flame across Britain burned a mixture of propane and butane.

Propane and butane are members of the same homologous series.

State what is meant by the term homologous series.   **1**

(b)   Natural gas, which is mainly methane, was used to fuel the flame in the Olympic cauldron.

(i)   Draw a diagram to show how **all** the outer electrons are arranged in a molecule of methane, CH_4.   **1**

(ii)   Methane is a covalent molecular substance. It has a low boiling point and is a gas at room temperature.

Explain why methane is a gas at room temperature.   **2**

[Turn over

MARKS | DO NOT WRITE IN THIS MARGIN

8. Car manufacturers have developed vehicles that use ethanol as fuel.

(a) The structure of ethanol is shown below.

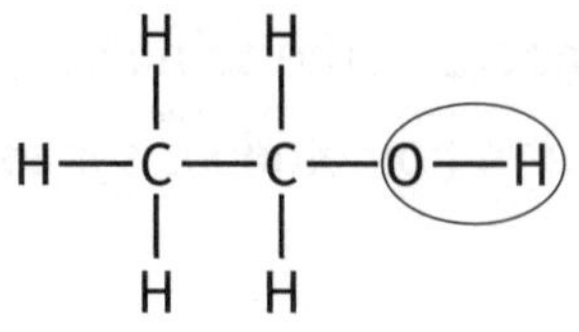

Name the functional group circled in the diagram.

1

(b) Name the two substances produced when ethanol burns in a plentiful supply of oxygen.

1

(c) Ethanol can be produced from ethene as shown.

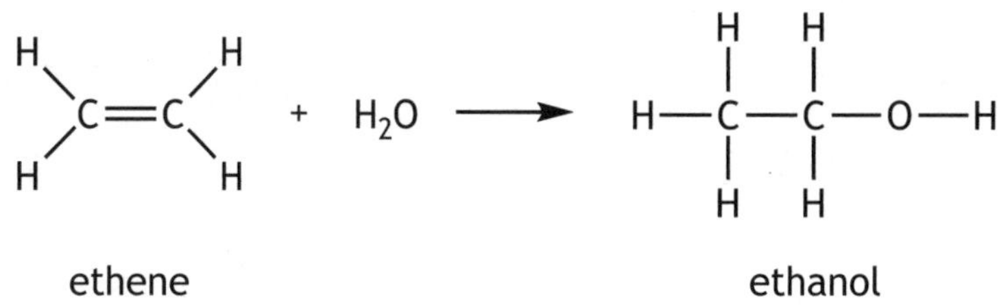

ethene   ethanol

(i) Name the **type** of chemical reaction taking place.

1

MARKS | DO NOT WRITE IN THIS MARGIN

8. (c) (continued)

(ii) Draw a structural formula for a product of the following reaction.　　**1**

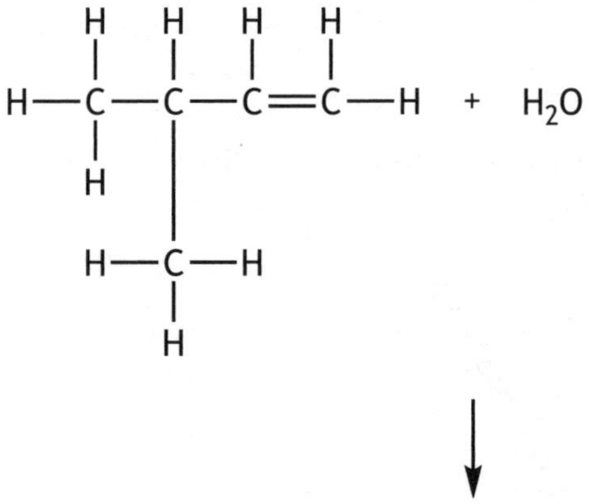

$$
\begin{array}{c}
\text{H}\quad\text{H}\quad\text{H}\quad\text{H}\\
||||\\
\text{H}-\text{C}-\text{C}-\text{C}=\text{C}-\text{H}\quad+\quad H_2O\\
||\\
\text{H}|\\
\text{H}-\text{C}-\text{H}\\
|\\
\text{H}
\end{array}
$$

↓

(d) Ethanol can be used to produce ethanoic acid.

(i) Draw a structural formula for ethanoic acid.　　**1**

(ii) Name the family of compounds to which ethanoic acid belongs.　　**1**

[Turn over

MARKS | DO NOT WRITE IN THIS MARGIN

9. Alkanes burn, releasing heat energy.

(a) State the term used to describe all chemical reactions that release heat energy.

1

(b) A student investigated the amount of energy released when an alkane burns using the apparatus shown.

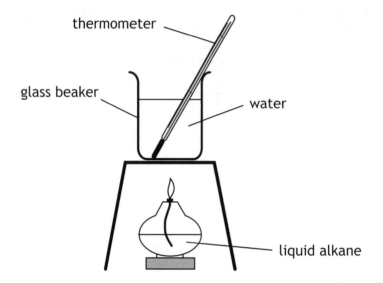

The student recorded the following data.

Mass of alkane burned	1 g
Volume of water	200 cm^3
Initial temperature of water	15 °C
Final temperature of water	55 °C

(i) Calculate the energy released, in kJ.

Show your working clearly.

3

MARKS | DO NOT WRITE IN THIS MARGIN

9. (b) (continued)

(ii) Suggest **one** improvement to the student's investigation.   1

(c) The table gives information about the amount of energy released when one mole of some alkanes are burned.

Name of alkane	Energy released when one mole of alkane is burned (kJ)
methane	891
ethane	1561
propane	2219
butane	2878

(i) Write a statement linking the amount of energy released to the number of carbon atoms in the alkane molecule.   1

(ii) Predict the amount of heat released, in kJ, when one mole of pentane is burned.   1

[Turn over

MARKS | DO NOT WRITE IN THIS MARGIN

10. Essential oils can be extracted from plants and used in perfumes and food flavourings.

    (a) Essential oils contain compounds made up of a number of isoprene molecules joined together.

        The shortened structural formula for isoprene is $CH_2C(CH_3)CHCH_2$.

        Draw the **full** structural formula for isoprene.　　　1

    (b) Essential oils can be extracted from the zest of lemons in the laboratory by steam distillation.

        The process involves heating up water in a boiling tube until it boils. The steam produced then passes over the lemon zest which is separated from the water by glass wool. As the steam passes over the lemon zest it carries essential oils into the delivery tube. The condensed liquids (essential oils and water) are collected in a test tube placed in a cold water bath.

        Complete the diagram to show the apparatus needed to collect the essential oils.　　　1

        (An additional diagram, if required, can be found on *Page twenty-nine*.)

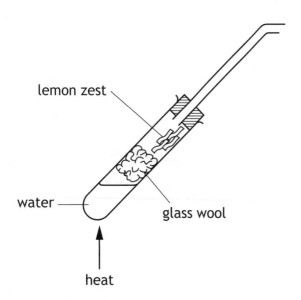

MARKS | DO NOT WRITE IN THIS MARGIN

10. **(continued)**

(c) Limonene, $C_{10}H_{16}$, is a compound found in lemon zest.

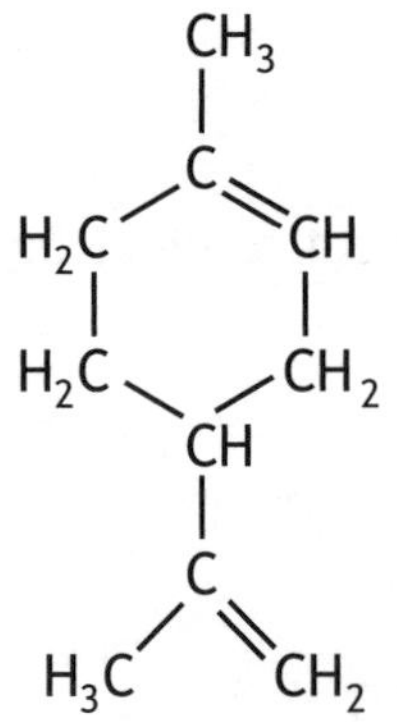

Write the molecular formula for the product formed when limonene reacts completely with bromine solution.

1

[Turn over

MARKS | DO NOT WRITE IN THIS MARGIN

11. Metals can be extracted from metal compounds by electrolysis.

(a) During electrolysis, metal ions are changed to metal atoms.

Name this type of chemical reaction.   **1**

(b) A student set up the following experiment to electrolyse copper(II) chloride solution.

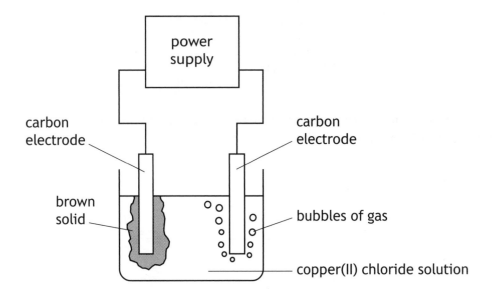

(i) Name the type of power supply that **must** be used to electrolyse the solution.   **1**

(ii) Complete the table by adding the charge on each electrode.   **1**

Observation at the _____ electrode	Observation at the _____ electrode
brown solid formed	bubbles of gas

MARKS | DO NOT WRITE IN THIS MARGIN

12. Urea, H_2NCONH_2, can be used as a fertiliser.

(a) Calculate the percentage of nitrogen in urea.   **3**

(b) Other nitrogen based fertilisers can be produced from ammonia.
Ammonia is produced in an industrial process using a catalyst.

$$N_2(g) + 3H_2(g) \rightleftharpoons 2NH_3(g)$$

(i) Name the industrial process that produces ammonia.   **1**

(ii) Suggest why a catalyst may be used in an industrial process.   **1**

(c) In another industrial process, ammonia is used to produce nitric acid.
Name the catalyst used in this process.   **1**

[Turn over

MARKS | DO NOT WRITE IN THIS MARGIN

13. Vitamin C is found in fruits and vegetables.

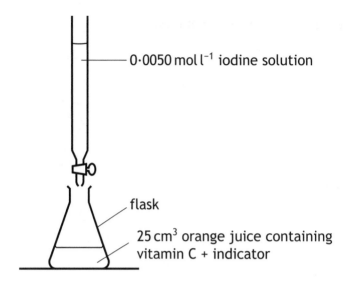

0·0050 mol l⁻¹ iodine solution

flask

25 cm³ orange juice containing vitamin C + indicator

Using iodine solution, a student carried out experiments to determine the concentration of vitamin C in orange juice.

The results of the experiments are shown.

Experiment	Initial volume of iodine solution (cm³)	Final volume of iodine solution (cm³)	Volume of iodine solution added (cm³)
1	1·2	18·0	16·8
2	18·0	33·9	15·9
3	0·5	16·6	16·1

(a) (i) Name the piece of apparatus used to measure the volume of iodine solution added to the orange juice.   1

   (ii) Calculate the average volume, in cm³, of iodine solution that should be used in calculating the concentration of vitamin C.   1

      Show your working clearly.

(b) Name the experimental method, carried out by the student, to accurately determine the concentration of vitamin C in the orange juice.   1

MARKS | DO NOT WRITE IN THIS MARGIN

14. In medicine, technetium-99m is injected into the body to detect damage to heart tissue.

It is a gamma-emitting radioisotope with a half-life of 6 hours.

(a) A sample of technetium-99m has a mass of 2 g.

Calculate the mass, in grams, of technetium-99m that would be left after 12 hours.   **2**

Show your working clearly.

(b) Suggest one reason why technetium-99m can be used safely in this way.   **1**

(c) Technetium-99m is formed when molybdenum-99 decays.

The decay equation is:

$$^{99}_{42}\text{Mo} \rightarrow {}^{99}_{43}\text{Tc} + \text{X}$$

Identify **X**.   **1**

[Turn over

MARKS | DO NOT WRITE IN THIS MARGIN

15. The concentration of chloride ions in water affects the ability of some plants to grow.

A student investigated the concentration of chloride ions in the water at various points along the river Tay.

The concentration of chloride ions in water can be determined by reacting the chloride ions with silver ions.

$$Ag^+(aq) \ + \ Cl^-(aq) \ \rightarrow \ AgCl(s)$$

A $20\,cm^3$ water sample gave a precipitate of silver chloride with a mass of $1.435\,g$.

(a) Calculate the number of moles of silver chloride, $AgCl$, present in this sample.

2

(b) Using your answer to part (a), calculate the concentration, in $mol\,l^{-1}$, of chloride ions in this sample.

2

16. Nitrogen, phosphorus and potassium are elements essential for plant growth.

   A student was asked to prepare a dry sample of a compound which contained **two** of these elements.

   The student was given access to laboratory equipment and the following chemicals.

Chemical	Formula
ammonium hydroxide	NH_4OH
magnesium nitrate	$Mg(NO_3)_2$
nitric acid	HNO_3
phosphoric acid	H_3PO_4
potassium carbonate	K_2CO_3
potassium hydroxide	KOH
sodium hydroxide	$NaOH$
sulfuric acid	H_2SO_4
water	H_2O

   Using your knowledge of chemistry, comment on how the student could prepare their dry sample.

   3

[END OF SPECIMEN QUESTION PAPER]

MARKS

ADDITIONAL SPACE FOR ANSWERS

Additional graph paper for Question 4 (b) (i)

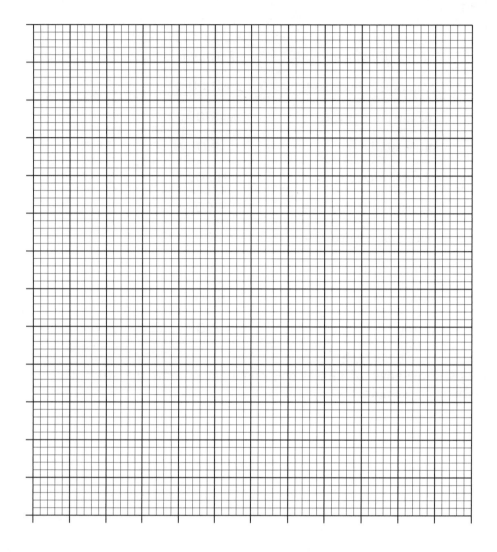

ADDITIONAL SPACE FOR ANSWERS

Additional diagram for Question 10 (b)

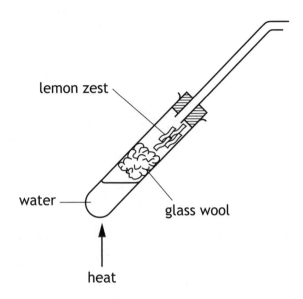

MARKS | DO NOT WRITE IN THIS MARGIN

ADDITIONAL SPACE FOR ANSWERS AND ROUGH WORK

NATIONAL 5

2018

National Qualifications 2018

X813/75/02

Chemistry
Section 1—Questions

MONDAY, 21 MAY

1:00 PM – 3:30 PM

Instructions for completion of Section 1 are given on *Page two* of your question and answer booklet X813/75/01.

Record your answers on the answer grid on *Page three* of your question and answer booklet.

You may refer to the Chemistry Data Booklet for National 5.

Before leaving the examination room you must give your question and answer booklet to the Invigilator; if you do not, you may lose all the marks for this paper.

SECTION 1 — 25 marks
Attempt ALL questions

1. Which of the following changes would **not** speed up a chemical reaction?

   A   Increasing the particle size

   B   Increasing the temperature

   C   Increasing the concentration

   D   Addition of a catalyst

2. Which line in the table identifies the correct location of a proton and an electron in an atom?

	Proton	Electron
A	inside the nucleus	inside the nucleus
B	inside the nucleus	outside the nucleus
C	outside the nucleus	outside the nucleus
D	outside the nucleus	inside the nucleus

3. Which of the following elements does **not** exist as diatomic molecules?

   A   Oxygen

   B   Helium

   C   Bromine

   D   Hydrogen

4. The shapes of some molecules are shown below.

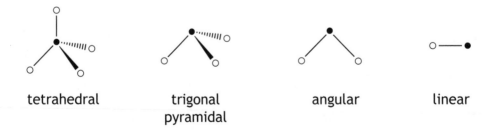

tetrahedral   trigonal   angular   linear
                     pyramidal

The shape of a molecule of hydrogen bromide is likely to be

   A   tetrahedral

   B   trigonal pyramidal

   C   angular

   D   linear.

5. Which of the following elements forms an ion with a single positive charge and an electron arrangement of 2,8?

You may wish to use the data booklet to help you.

A   Sodium

B   Magnesium

C   Fluorine

D   Neon

6. Which line in the table shows the properties of a covalent network compound?

	Melting point (°C)	Boiling point (°C)	Conducts electricity	
			Solid	Liquid
A	−127	−100	no	no
B	795	1410	no	yes
C	30	2204	yes	yes
D	2700	3350	no	no

7. 0·1 mol of sodium hydroxide was dissolved in water and the solution made up to $250 \, cm^3$.

What is the concentration, in $mol \, l^{-1}$, of the sodium hydroxide solution?

A   0·0004

B   0·025

C   0·4

D   2·5

8. An alkaline solution contains

A   only hydroxide ions

B   more hydroxide ions than hydrogen ions

C   more hydrogen ions than hydroxide ions

D   equal numbers of hydrogen ions and hydroxide ions.

[Turn over

9. A student made some statements about the effect of adding water to an acidic solution. Identify the correct statement.

   A   The pH of the solution will remain the same.

   B   The pH of the solution will decrease.

   C   The hydrogen ion concentration will decrease.

   D   The hydrogen ion concentration will increase.

10. The shortened structural formula for a compound is

$$CH_3CH_2CH(CH_3)CH(CH_3)CH_2CH_2CH_3$$

   Which of the following is another way of representing this structure?

   A

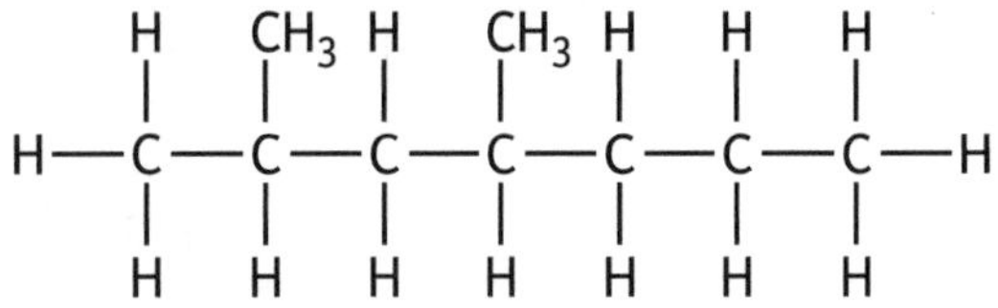

   B

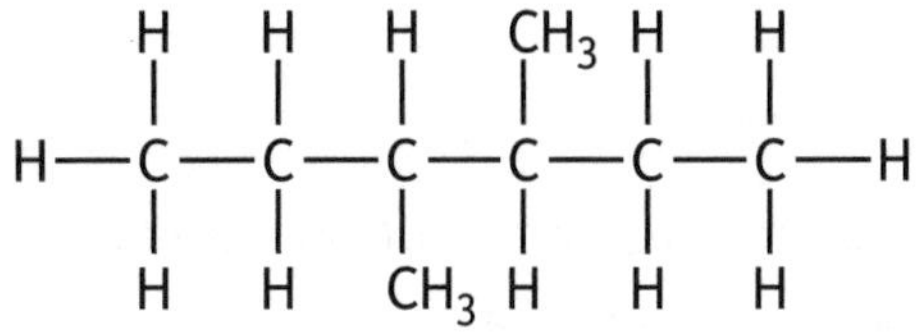

   C

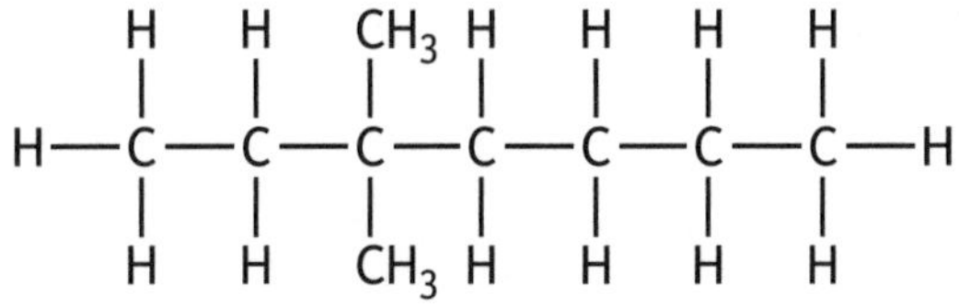

   D

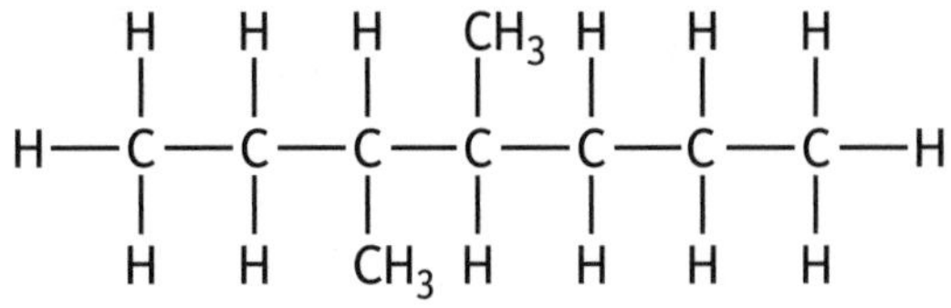

11. Identify which of the following is an isomer of

```
        H    H    H    H    H
        |    |    |    |    |
    H — C — C — C — C — C — H
        |    |    |    |    |
        H    |    H    H    H
        H — C — H
             |
             H
```

A
```
            H     H
             \   /
        H     C      H
         \   / \    /
      H — C       C — H
      H — C       C — H
         /   \   /   \
        H     C      H
             / \
            H   H
```

B
```
        H    H    H    H    H
        |    |    |    |    |
    H — C — C — C — C — C — H
        |    |    |    |    |
        H    H    H    |    H
                  H — C — H
                       |
                       H
```

C
```
        H    H    H    H    H
        |    |    |    |    |
    H — C — C — C — C — C — H
        |    |    |    |    |
        H    H    |    H    H
             H — C — H
                  |
                  H
```

D
```
        H    H    H    H
        |    |    |    |
    H — C — C — C — C — H
        |    |    |    |
        H    H    |    H
             H — C — H
                  |
                  H
```

12. Which of the following reactions takes place when an alcohol is formed from an alkene?

    A Hydrogenation

    B Combustion

    C Hydration

    D Reduction

13.

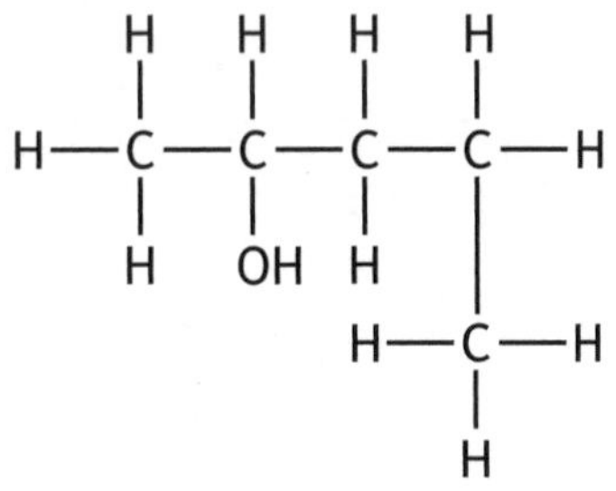

The systematic name for the above compound is

    A pentan-2-ol

    B pentan-4-ol

    C 1-methylbutan-3-ol

    D 4-methylbutan-2-ol.

14. Which of the following alcohols is the least soluble in water?

    A Butan-1-ol

    B Hexan-1-ol

    C Pentan-1-ol

    D Propan-1-ol

15. A student set up an experiment to determine the quantity of energy released when a hydrocarbon burns.

Which of the following diagrams shows the apparatus which would produce the most accurate result?

A   thermometer

glass beaker

draught shield

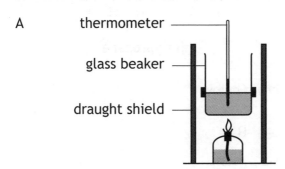

B   thermometer

metal can

draught shield

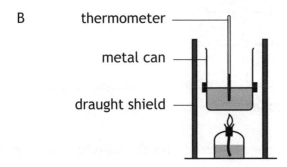

C   thermometer

metal can

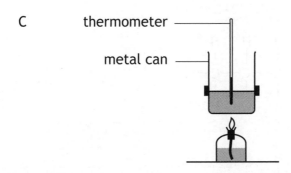

D   thermometer

glass beaker

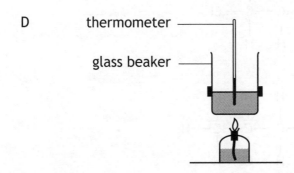

[Turn over

16.   The ether, 1-ethoxypropane, can be made by the Williamson reaction.

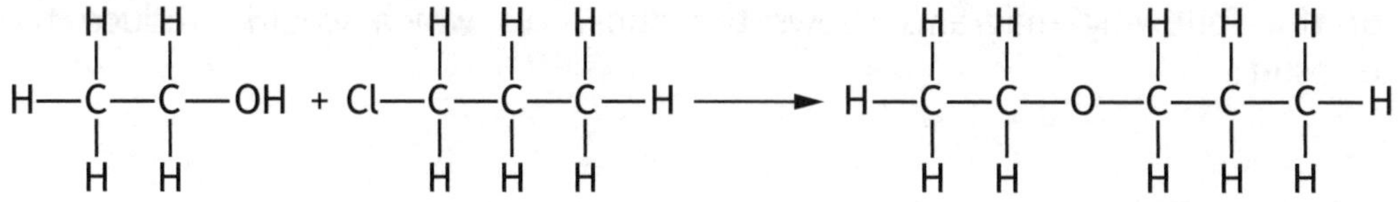

ethanol   1-chloropropane   1-ethoxypropane

The structural formula for another ether is shown below.

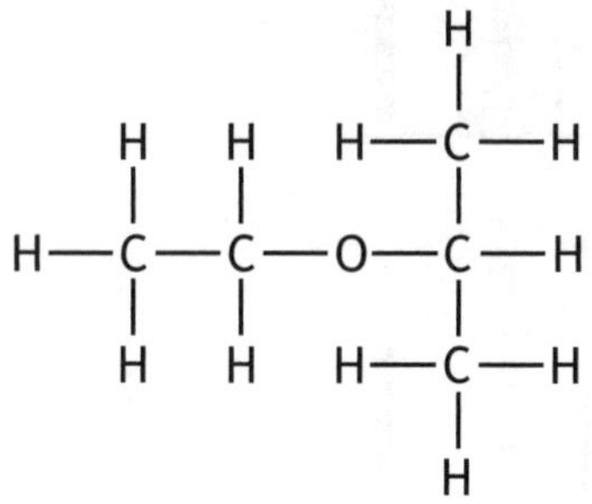

2-ethoxypropane

Which of the following pairs of compounds would react together to produce 2-ethoxypropane?

A

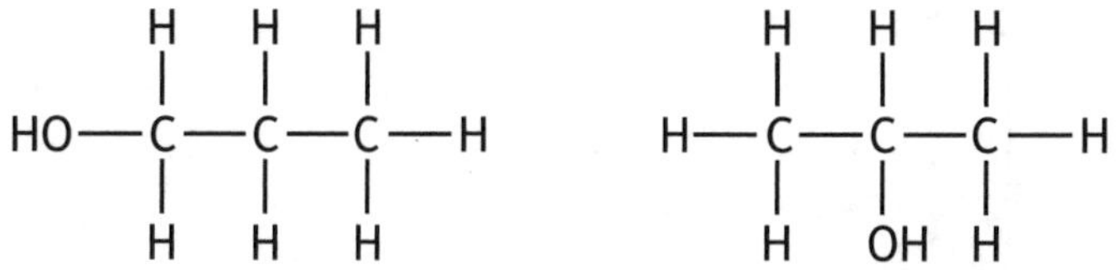

B

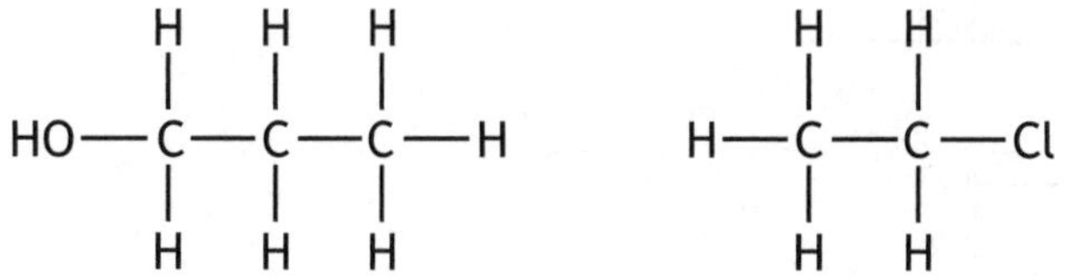

C

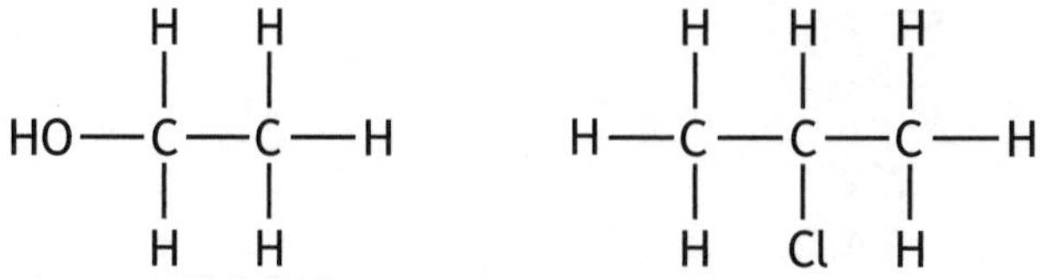

D

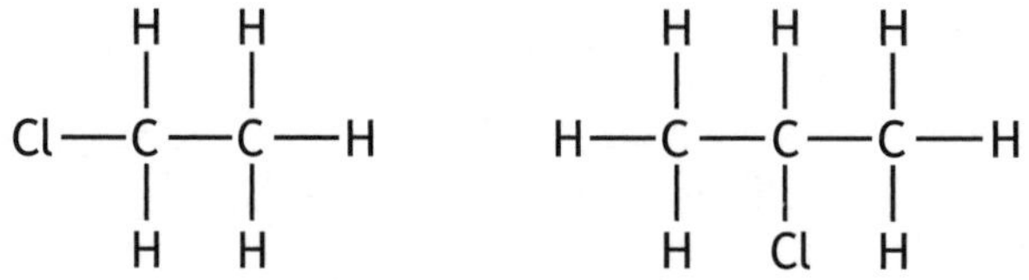

17. Information about the reactions of four different metals, **W**, **X**, **Y** and **Z** is given in the table.

Metal	Reaction with dilute acid	Reaction with water
W	moderate reaction	no reaction
X	fast reaction	slow reaction
Y	slow reaction	no reaction
Z	fast reaction	no reaction

The order of reactivity of the metals, starting with the most reactive is

A   X, Z, W, Y

B   Y, W, Z, X

C   Z, X, W, Y

D   Y, W, X, Z.

18. The ion-electron equations for the oxidation and reduction steps in the reaction between hydrogen and oxygen are given below.

$$H_2(g) \rightarrow 2H^+(aq) + 2e^-$$

$$2H_2O(\ell) + O_2(g) + 4e^- \rightarrow 4OH^-(aq)$$

The redox equation for the overall reaction is

A $H_2(g) + 2H_2O(\ell) + O_2(g) + 4e^- \rightarrow 2H^+(aq) + 4OH^-(aq) + 2e^-$

B $2H_2(g) + 2H_2O(\ell) + O_2(g) \rightarrow 4H^+(aq) + 4OH^-(aq)$

C $H_2(g) + 2H_2O(\ell) + O_2(g) \rightarrow 2H^+(aq) + 4OH^-(aq)$

D $2H_2(g) + 2H_2O(\ell) + O_2(g) + 4e^- \rightarrow 4H^+(aq) + 4OH^-(aq) + 4e^-$

[Turn over

19. Which of the following metals, when connected to lead in a cell, would produce the highest reading on the voltmeter?

    You may wish to use the data booklet to help you.

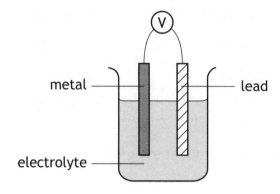

    A   Zinc

    B   Tin

    C   Nickel

    D   Lead

20. Which of the following salts would **not** be used as a fertiliser?

    A   Ammonium chloride

    B   Ammonium phosphate

    C   Sodium chloride

    D   Sodium phosphate

21. Which metal is used as the catalyst in the industrial manufacture of ammonia?

    A   Nickel

    B   Platinum

    C   Iron

    D   Rhodium

22. The diagram shows the path of two different types of radiation as they pass through an electric field.

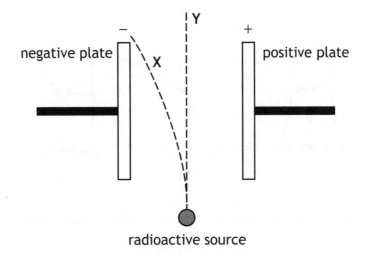

Which line in the table correctly identifies the types of radiation which follow paths X and Y?

	Path X	Path Y
A	alpha	beta
B	beta	alpha
C	beta	gamma
D	alpha	gamma

23. Metallic bonding is a force of attraction between

    A   a shared pair of electrons and two nuclei

    B   negative ions and delocalised electrons

    C   negative ions and positive ions

    D   positive ions and delocalised electrons.

[Turn over

24. $2K^+(aq) + 2I^-(aq) + Pb^{2+}(aq) + 2NO_3^-(aq) \rightarrow PbI_2(s) + 2K^+(aq) + 2NO_3^-(aq)$

   The type of reaction represented by this equation is

   A   neutralisation

   B   precipitation

   C   addition

   D   redox.

25. A student prepared a sample of copper sulfate crystals by reacting excess copper carbonate with acid.

   X   Y   Z

   Which line in the table shows the correct order in which this experiment would be carried out?

   A   Y, X, Z

   B   X, Y, Z

   C   Z, Y, X

   D   Y, Z, X

[END OF SECTION 1. NOW ATTEMPT THE QUESTIONS IN SECTION 2 OF YOUR QUESTION AND ANSWER BOOKLET]

FOR OFFICIAL USE

Mark

N5

National
Qualifications
2018

X813/75/01

**Chemistry
Section 1—Answer Grid
and Section 2**

MONDAY, 21 MAY

1:00 PM – 3:30 PM

Fill in these boxes and read what is printed below.

Full name of centre

Town

Forename(s)

Surname

Number of seat

Date of birth

Day	Month	Year		Scottish candidate number

Total marks— 100

SECTION 1— 25 marks

Attempt ALL questions.

Instructions for the completion of Section 1 are given on *Page two*.

SECTION 2— 75 marks

Attempt ALL questions.

You may refer to the Chemistry Data Booklet for National 5.

Write your answers clearly in the spaces provided in this booklet. Additional space for answers and rough work is provided at the end of this booklet. If you use this space you must clearly identify the question number you are attempting. Any rough work must be written in this booklet. You should score through your rough work when you have written your final copy.

Use **blue** or **black** ink.

Before leaving the examination room you must give this booklet to the Invigilator; if you do not, you may lose all the marks for this paper.

SECTION 1 — 25 marks

The questions for Section 1 are contained in the question paper X813/75/02.

Read these and record your answers on the answer grid on *Page three* opposite.

Use **blue** or **black** ink. Do NOT use gel pens or pencil.

1. The answer to each question is **either** A, B, C or D. Decide what your answer is, then fill in the appropriate bubble (see sample question below).

2. There is **only one correct** answer to each question.

3. Any rough working should be done on the additional space for answers and rough work at the end of this booklet.

Sample question

To show that the ink in a ball-pen consists of a mixture of dyes, the method of separation would be

 A fractional distillation

 B chromatography

 C fractional crystallisation

 D filtration.

The correct answer is **B** — chromatography. The answer **B** bubble has been clearly filled in (see below).

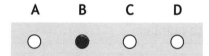

Changing an answer

If you decide to change your answer, cancel your first answer by putting a cross through it (see below) and fill in the answer you want. The answer below has been changed to **D**.

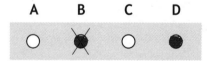

If you then decide to change back to an answer you have already scored out, put a tick (✓) to the **right** of the answer you want, as shown below:

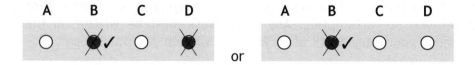

or

SECTION 1 — Answer Grid

	A	B	C	D
1	○	○	○	○
2	○	○	○	○
3	○	○	○	○
4	○	○	○	○
5	○	○	○	○
6	○	○	○	○
7	○	○	○	○
8	○	○	○	○
9	○	○	○	○
10	○	○	○	○
11	○	○	○	○
12	○	○	○	○
13	○	○	○	○
14	○	○	○	○
15	○	○	○	○
16	○	○	○	○
17	○	○	○	○
18	○	○	○	○
19	○	○	○	○
20	○	○	○	○
21	○	○	○	○
22	○	○	○	○
23	○	○	○	○
24	○	○	○	○
25	○	○	○	○

[BLANK PAGE]

DO NOT WRITE ON THIS PAGE

[Turn over for next question

DO NOT WRITE ON THIS PAGE

MARKS | DO NOT WRITE IN THIS MARGIN

SECTION 2 — 75 marks

Attempt ALL questions

1. A student monitored the rate of reaction between excess calcium carbonate and dilute hydrochloric acid, HCl, using a gas syringe to collect the gas produced.

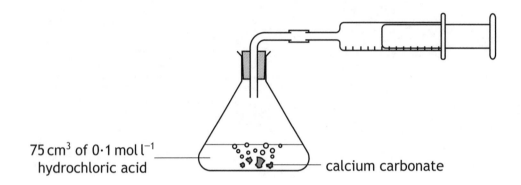

75 cm³ of 0·1 mol l⁻¹ hydrochloric acid

calcium carbonate

(a) Name the gas produced in this reaction. 1

(b) The student obtained the results shown.

Time (s)	0	10	20	40	50	60	70	80
Volume of gas (cm³)	0	48	62	74	77	79	80	80

(i) Calculate the average rate of reaction between 20 and 50 seconds. 3
Your answer must include the appropriate unit.
Show your working clearly.

MARKS | DO NOT WRITE IN THIS MARGIN

1. (b) (continued)

    (ii) Draw a graph of the student's results.   **4**

        (Additional graph paper, if required, can be found on *Page thirty-three*).

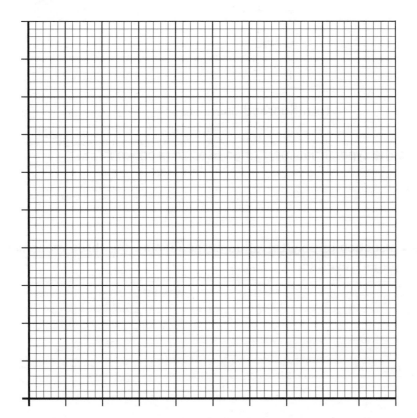

    (iii) **Using your graph**, estimate the volume of gas, in cm^3, produced at 30 seconds.   **1**

(c) The student carried out a similar experiment using $75\,cm^3$ of $0\cdot1\,mol\,l^{-1}$ sulfuric acid, H_2SO_4(aq).

The total volume of gas collected was $160\,cm^3$.

Explain why there was a greater volume of gas produced.   **1**

MARKS | DO NOT WRITE IN THIS MARGIN

2. The retractable roof on Centre Court at Wimbledon Tennis Club is made of the polymer poly(tetrafluoroethene), PTFE.

   (a) The monomer used to produce PTFE has the following structure.

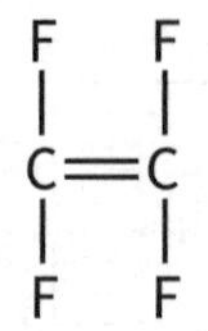

$$\begin{array}{ccc} F & & F \\ | & & | \\ C & = & C \\ | & & | \\ F & & F \end{array}$$

   tetrafluoroethene

   (i) Name the type of polymerisation used to produce PTFE.   1

   (ii) Draw a section of poly(tetrafluoroethene) showing three monomer units joined together.   1

   (b) The roof of the O_2 Arena in London is made from a co-polymer.

   A co-polymer is formed when two different monomers polymerise.

   The repeating unit of the co-polymer is shown.

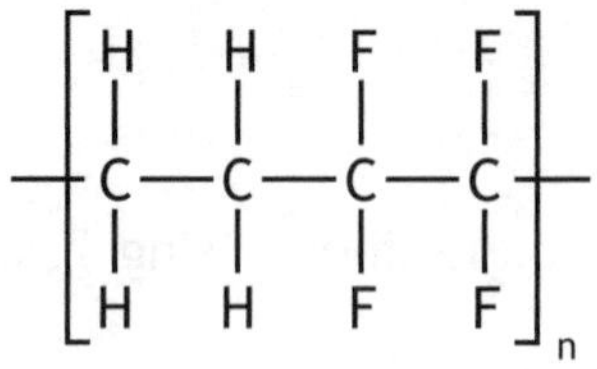

$$\left[\begin{array}{cccc} H & H & F & F \\ | & | & | & | \\ C & C & C & C \\ | & | & | & | \\ H & H & F & F \end{array}\right]_n$$

   One of the monomers in this co-polymer is tetrafluoroethene.

   Draw the full structural formula for the other monomer.   1

MARKS | DO NOT WRITE IN THIS MARGIN

3. Coal is a fuel that contains carbon. Different types of coal contain different percentages of carbon.

Heat content is a measure of how much heat energy is released when coal is burned.

(a) The table gives information about types of coal.

Type of coal	Percentage of carbon	Average heat content ($kJ\,kg^{-1}$)
Anthracite	86–98	32 500
Bituminous	45–85	27 850
Sub-bituminous	35–44	25 550
Lignite	25–34	13 950

Describe how the percentage of carbon in coal affects the average heat content.

1

(b) Iron pyrite, FeS_2, is an impurity found in coal.

Calculate the percentage of iron in iron pyrite.

Show your working clearly.

3

[Turn over

MARKS | DO NOT WRITE IN THIS MARGIN

4. During the FIFA World Cup, referees will spray foam onto the pitch to ensure players stand the correct distance from the ball when a free kick is taken. The foam contains a hydrocarbon mixture of isobutane, butane and propane.

(a) Name the elements present in a hydrocarbon.

1

(b) The full structural formula for isobutane is

$$
\begin{array}{c}
\quad\quad\;\; H \\
\quad\quad\;\; | \\
H-C-H \\
H \quad | \quad H \\
| \quad\;\; | \quad\;\; | \\
H-C-C-C-H \\
| \quad\;\; | \quad\;\; | \\
H \quad H \quad H
\end{array}
$$

Write the systematic name for isobutane.

1

(c) Alkanes have different physical properties.

The table gives some information about isobutane and butane.

Alkane	Boiling point (°C)
isobutane	−12
butane	−1

Circle the correct words to complete the sentence.

1

Compared to isobutane, butane has a higher boiling point

as it contains { weaker / stronger } { covalent bonds / intermolecular forces }.

MARKS | DO NOT WRITE IN THIS MARGIN

4.   (continued)

(d)   The table shows the boiling points of some alkanes.

Alkane	Boiling point (°C)
pentane	36
hexane	69
heptane	98
octane	126
nonane	

Predict the boiling point, in °C, of nonane, C_9H_{20}.   1

[Turn over

MARKS | DO NOT WRITE IN THIS MARGIN

5. Read the passage and answer the questions that follow.

The Chemistry within Airbags

Airbags, an important safety feature in cars, inflate rapidly on collision. Inside the airbag is a gas generator containing a mixture of sodium azide (NaN_3), potassium nitrate and silicon dioxide.

When a car is involved in a collision, a series of three chemical reactions takes place.

In the first reaction, electrical energy causes sodium azide to decompose producing sodium metal and nitrogen gas. The nitrogen gas that is generated fills the airbag.

In the second reaction, the sodium reacts with potassium nitrate producing more nitrogen gas, sodium oxide and potassium oxide.

In the final reaction, the metal oxides react with silicon dioxide to produce silicate fibres, which are harmless and stable.

This process, from the initial impact of the crash to full inflation of the airbag, takes a fraction of a second.

(a) Name the three chemicals found inside the gas generator before any chemical reactions take place.   1

(b) Name the compound produced in the second reaction which would give a lilac flame colour.   1

You may wish to use the data booklet to help you.

(c) Write the formula for the compound which reacts with the metal oxides in the final reaction.   1

MARKS | DO NOT WRITE IN THIS MARGIN

5. **(continued)**

(d) The graph below gives information on the volume of nitrogen gas produced by the gas generator.

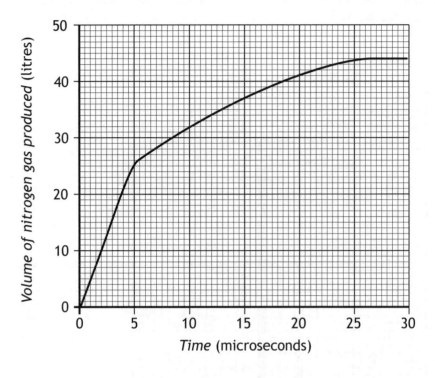

State the total volume, in litres, of nitrogen gas produced. 1

[Turn over

6. Scientists use an instrument called a mass spectrometer to determine the number of isotopes and the percentage of each isotope in a sample of an element.

   (a) When a sample of boron is passed through a mass spectrometer the following graph is obtained.

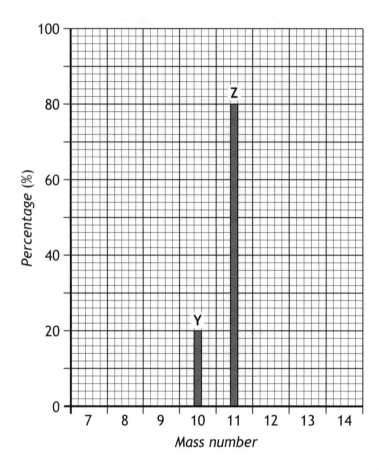

   (i) State the number of isotopes present in this sample of boron.   1

MARKS | DO NOT WRITE IN THIS MARGIN

6. (a) (continued)

    (ii) The relative atomic mass can be calculated using:

$$\frac{(\text{mass of isotope } \mathbf{Y} \times \% \text{ of } \mathbf{Y}) + (\text{mass of isotope } \mathbf{Z} \times \% \text{ of } \mathbf{Z})}{100}$$

    Using the information from the graph, calculate the relative atomic mass of the sample of boron.　2

    Show your working clearly.

(b) Carbon also has more than one isotope.

    The nuclide notation for an isotope of carbon can be represented as

$$^{12}_{6}\text{C}$$

    Write the nuclide notation for the isotope of carbon with 8 neutrons.　1

[Turn over

MARKS | DO NOT WRITE IN THIS MARGIN

7. Strontium chloride, which is an ionic compound, is used in toothpaste to reduce tooth sensitivity.

(a) State the term used to describe the structure of solid strontium chloride.   **1**

(b) A sample of strontium chloride was electrolysed.

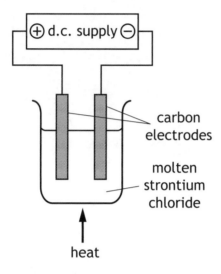

(i) State why ionic compounds, like strontium chloride, conduct electricity when molten.   **1**

(ii) During electrolysis, chloride ions lose electrons to form chlorine gas.

Name the type of chemical reaction taking place.   **1**

(iii) Explain why a d.c. supply **must** be used.   **1**

MARKS | DO NOT WRITE IN THIS MARGIN

8. Water is one of the most versatile of all chemicals and features in many chemical reactions and processes.

Using your knowledge of chemistry, comment on the chemistry of water.　　**3**

[Turn over

MARKS | DO NOT WRITE IN THIS MARGIN

9. Olive oil, which can be used in cooking, is a mixture of unsaturated molecules.

   (a) (i) State what is meant by the term unsaturated.   **1**

       (ii) Describe the chemical test, including the result, that can be used to show that olive oil is unsaturated.   **1**

   (b) When frying food, it is recommended that the oil is heated before food is added.

   The table gives information about olive oil used to fry food.

Specific heat capacity of olive oil	$1.97\,kJ\,kg^{-1}\,°C^{-1}$
Initial temperature of olive oil	$20\,°C$
Mass of olive oil heated	$1500\,g$

   Calculate the energy, in kJ, required to increase the temperature of the olive oil to $180\,°C$.   **3**

   Show your working clearly.

MARKS | DO NOT WRITE IN THIS MARGIN

10. Ammonia is made industrially by reacting nitrogen with hydrogen.

(a) The equation for this reaction is

$$N_2 \quad + \quad H_2 \quad \rightleftharpoons \quad NH_3$$

    (i) Balance the equation above.   **1**

    (ii) In the equation the symbol $\rightleftharpoons$ is used.

         State what this indicates about the reaction.   **1**

(b) Draw a diagram, showing **all** outer electrons, to represent a molecule of ammonia, NH_3.   **1**

(c) In industry, ammonia can be converted into nitric acid.

Name this industrial process.   **1**

(d) Ammonia reacts with nitric acid to produce a salt.

Name the salt produced in this reaction.   **1**

MARKS | DO NOT WRITE IN THIS MARGIN

11. A student set up the following cell.

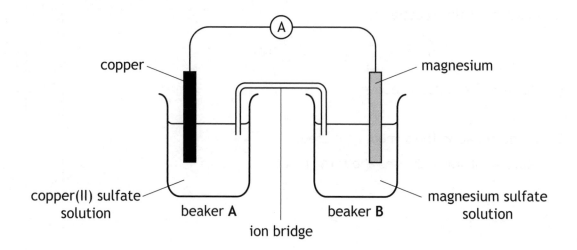

copper

magnesium

copper(II) sulfate solution

beaker **A**

beaker **B**

magnesium sulfate solution

ion bridge

(a) **On the diagram**, draw an arrow to show the path and direction of electron flow.

You may wish to use the data booklet to help you.

1

(b) Explain why an ion bridge is used to link the beakers.

1

(c) In this reaction, the copper ions are reduced.

Write the ion-electron equation for the reduction of copper(II) ions.

You may wish to use the data booklet to help you.

1

MARKS | DO NOT WRITE IN THIS MARGIN

11. (continued)

(d) Other magnesium compounds could be used in place of magnesium sulfate when making this type of cell.

Suggest why magnesium phosphate would **not** be suitable.   1

You may wish to use the data booklet to help you.

[Turn over

MARKS | DO NOT WRITE IN THIS MARGIN

12. Thallium-204 decays by emitting beta particles and can be used in industry to measure the thickness of paper.

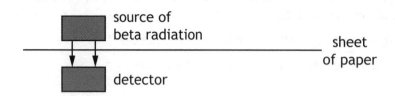

source of beta radiation

sheet of paper

detector

(a) Suggest a reason why a radioisotope which emits alpha particles is not suitable for this purpose.

1

(b) A paper manufacturer found a thallium-204 source had only $\frac{1}{16}$ of its original activity.

The half-life of thallium-204 is 3·7 years.

Calculate the age, in years, of the source.

2

Show your working clearly.

(c) Circle the correct words to complete the sentence.

1

When an atom emits a beta particle,

the atomic number of the atom $\left\{ \begin{array}{l} \text{increases} \\ \text{decreases} \\ \text{stays the same} \end{array} \right\}$ and

the mass number $\left\{ \begin{array}{l} \text{increases} \\ \text{decreases} \\ \text{stays the same} \end{array} \right\}$.

MARKS | DO NOT WRITE IN THIS MARGIN

13. Malic acid is a carboxylic acid found in some fruits.

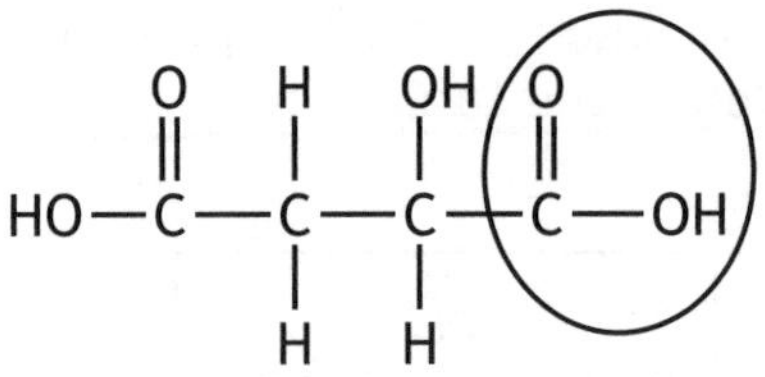

(a)   (i)   Name the functional group circled in the diagram above.

1

     (ii)   Calculate the mass, in grams, of 1 mole of malic acid.

1

[Turn over

MARKS | DO NOT WRITE IN THIS MARGIN

13. **(continued)**

(b) Carboxylic acids can contain a halogen atom. The pH of $1\,mol\,l^{-1}$ solutions of some of these acids are given in the table.

Carboxylic acid	pH
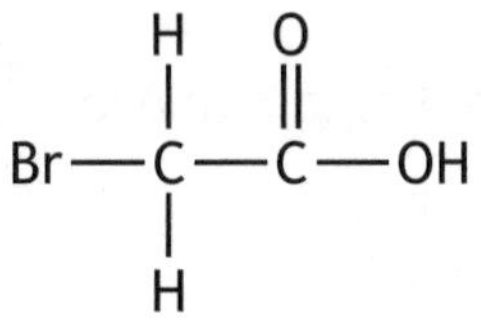 Br—C(H)(H)—C(=O)—OH	1·45
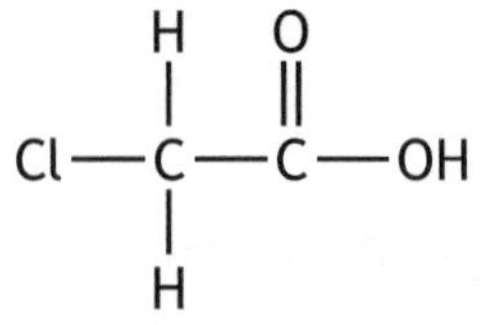 Cl—C(H)(H)—C(=O)—OH	1·42
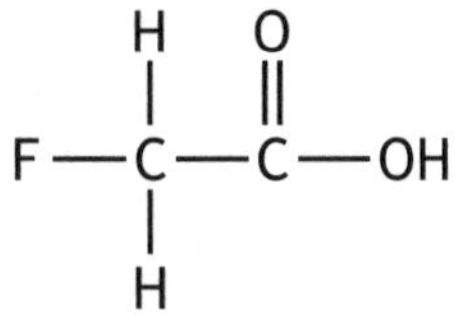 F—C(H)(H)—C(=O)—OH	1·33
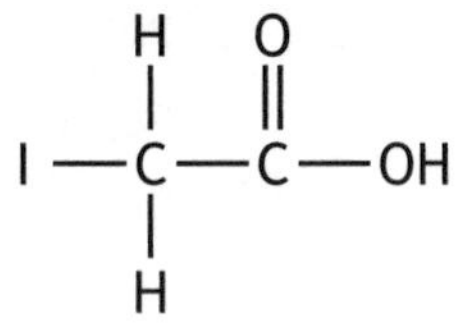 I—C(H)(H)—C(=O)—OH	1·55

Describe how the acidity of the carboxylic acid is related to the position of the halogen in group 7 of the periodic table.

1

MARKS | DO NOT WRITE IN THIS MARGIN

13. (continued)

(c) The Jones oxidation reaction can be used to convert alcohols to carboxylic acids.

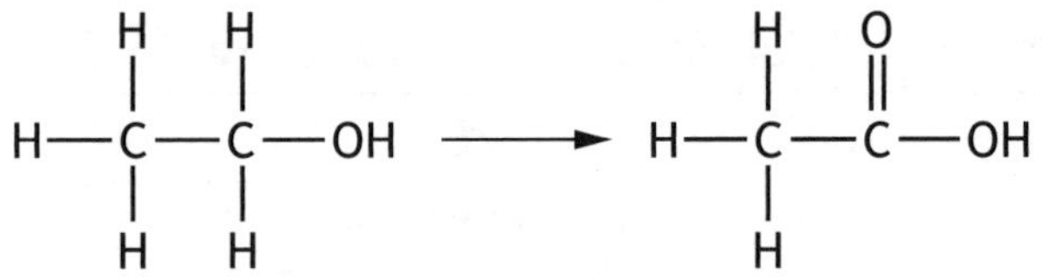

The following alcohol can also be converted to a carboxylic acid by the Jones oxidation reaction.

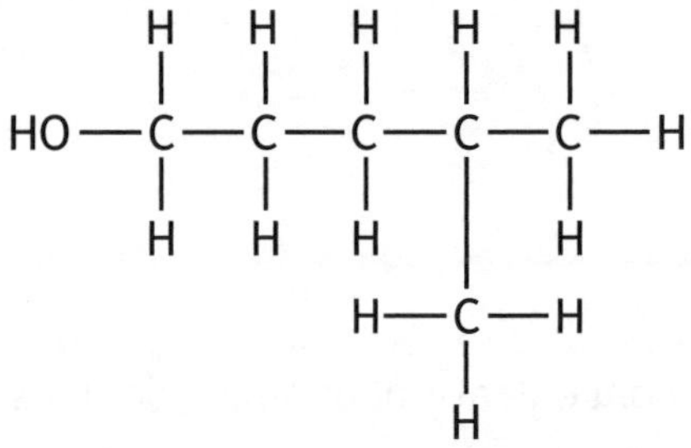

Draw a structural formula for the carboxylic acid produced in this reaction.

1

[Turn over

MARKS | DO NOT WRITE IN THIS MARGIN

14. Chloride ion concentrations greater than $0.25\,g\,l^{-1}$ can cause a noticeable taste in drinking water.

The table gives information about the chloride ion concentration in drinking water from different sources.

Source	Chloride ion concentration $(g\,l^{-1})$
A	0.26
B	0.28
C	0.24

(a) One of the sources provides drinking water that does **not** have a noticeable taste.

Identify this source.

1

(b) A student investigated the concentration of chloride ions in drinking water from another source.

Samples of water were titrated with silver nitrate solution.

An indicator was used to show when the end-point was reached.

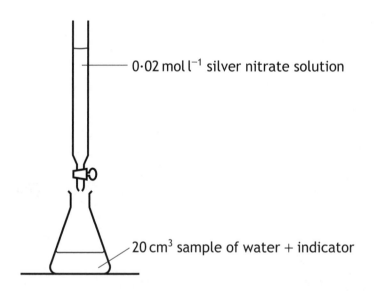

$0.02\,mol\,l^{-1}$ silver nitrate solution

$20\,cm^3$ sample of water + indicator

Titration	Volume of silver nitrate added (cm^3)
1	9.6
2	8.0
3	8.5
4	8.1

MARKS | DO NOT WRITE IN THIS MARGIN

14.   (b)   (continued)

(i)   Name the most appropriate piece of apparatus to measure $20\,cm^3$ samples of water into the flask.   **1**

(ii)   The average volume of silver nitrate that should be used to calculate the chloride ion concentration is $8 \cdot 05\,cm^3$.

Explain why only the results of titration **2** and titration **4** are used to calculate this average.   **1**

(iii)   Calculate the number of moles of silver nitrate in $8 \cdot 05\,cm^3$.   **1**

Show your working clearly.

[Turn over

15. Read the passage and answer the questions that follow.

The bizarre world of high pressure chemistry

What would happen if you put some sodium, normally a soft grey metal, under extremely high pressure?

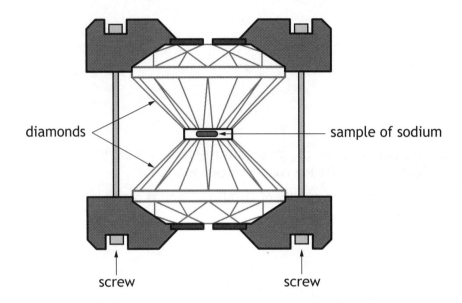

Researchers investigated this using a piece of apparatus called a diamond anvil cell. The diamond anvil cell contains two diamonds and as the screws are tightened, high pressure is created. The pressure between the diamonds can reach 1000 gigapascals, which is a pressure of 10 million atmospheres.

When sodium is squeezed to 190 gigapascals it loses an important property of metals and becomes an insulator. This shows that there is a change in the structure and bonding of sodium.

The diamond anvil cell also allows scientists to create new materials, including superconductors.

Scientists are studying what happens to the materials thought to be deep inside the Earth, where high pressure occurs naturally.

Using this technique to mirror what may happen to materials deep in the Earth, iron(III) oxide is found to decompose, releasing oxygen, and forming the very unusual Fe_5O_7.

Adapted from *The Catalyst*, Volume 27, Number 1, October 2016

MARKS | DO NOT WRITE IN THIS MARGIN

15. (continued)

(a) Name the piece of apparatus used by researchers to create high pressure.   1

(b) (i) Calculate the pressure, in atmospheres, when sodium is squeezed at 190 gigapascals.   1

(ii) Suggest what would be observed if this pressurised sodium was placed in the circuit below.   1

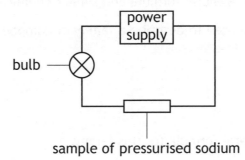

sample of pressurised sodium

(c) Write an equation, using symbols and formulae, to show the decomposition of iron(III) oxide, at high pressure.   1

There is no need to balance this equation.

MARKS | DO NOT WRITE IN THIS MARGIN

16. The thiols are a family of compounds containing carbon, hydrogen and sulfur.

Name	Full structural formula
methanethiol	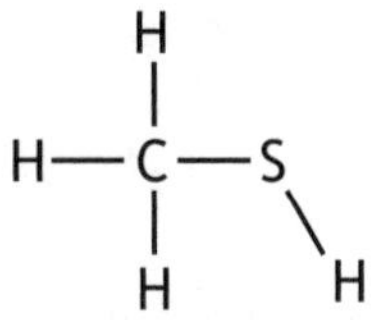
ethanethiol	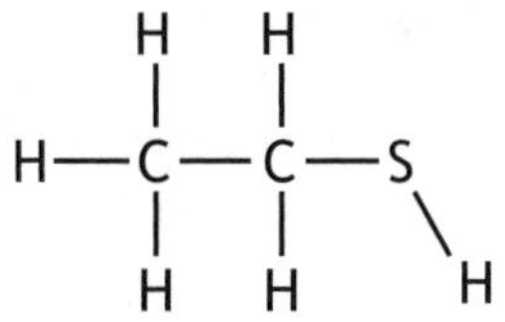
propanethiol	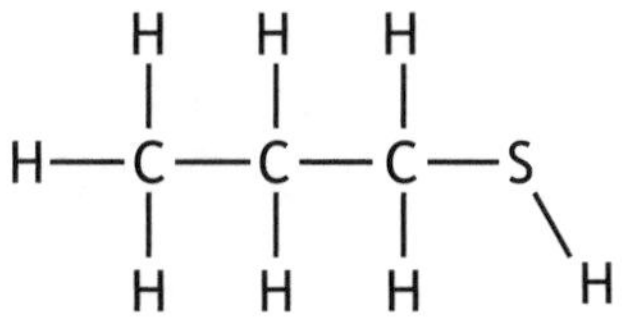

(a) Thiols have the same general formula and similar chemical properties.

(i) State the term used to describe a family of compounds such as the thiols.   1

(ii) Suggest a general formula for this family.   1

(b) Ethanethiol can react with oxygen as shown.

ethanethiol + oxygen → carbon dioxide + water + Y

Identify Y.   1

Page thirty

MARKS | DO NOT WRITE IN THIS MARGIN

16. (continued)

(c) Methanethiol, which smells like rotting cabbage, is added to natural gas to allow gas leaks to be detected.

It is prepared industrially by the reaction of methanol with hydrogen sulfide gas.

$$CH_3OH \ + \ H_2S \longrightarrow CH_3SH \ + \ H_2O$$

Calculate the mass of methanethiol, in grams, produced when 640 grams of methanol reacts completely with hydrogen sulfide.

Show your working clearly.

3

MARKS | DO NOT WRITE IN THIS MARGIN

17. Methacrylic acid is used to make methacrylates which are used in Shellac nail polish.

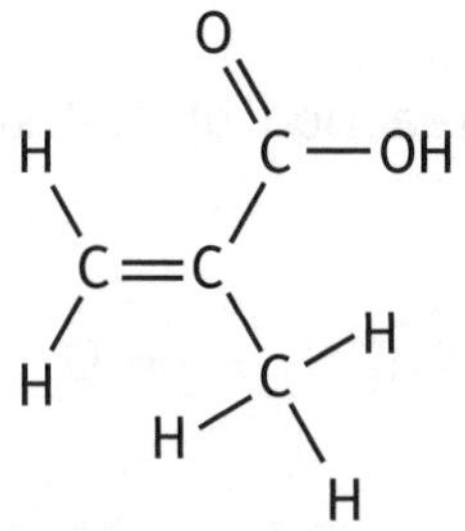

methacrylic acid

Using your knowledge of chemistry, comment on the chemistry of methacrylic acid.

3

[END OF QUESTION PAPER]

ADDITIONAL SPACE FOR ANSWERS AND ROUGH WORKING

Additional graph paper for Question 1 (b) (ii)

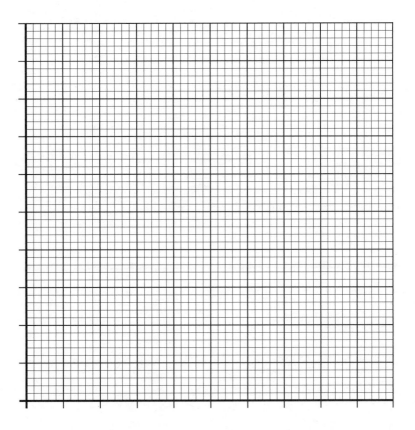

MARKS | DO NOT WRITE IN THIS MARGIN

ADDITIONAL SPACE FOR ANSWERS AND ROUGH WORKING

NATIONAL 5

Answers

NATIONAL 5 CHEMISTRY 2017

Section 1

Question	Response
1.	D
2.	C
3.	A
4.	D
5.	A
6.	B
7.	C
8.	B
9.	A
10.	A
11.	D
12.	C
13.	C
14.	B
15.	D
16.	C
17.	A
18.	B
19.	A
20.	C

Section 2

1. (a) Isotope(s)

   (b) Different numbers of neutrons

   or

   The atoms have 18, 20 or 22 neutrons

   (c) 36

   or

   $^{36}_{18}Ar$

   or

   ^{36}Ar

2. (a) Carbon nanotube

   or

   Nanotube

   (b) Lithium or Li

   (c) 20·5 with no working

   21 with correct working

   Partial marking:
   Demonstration of the correct use of the relationship concept
   i.e. 41/2.   (1)

   or

   41/1 = 41   (1)

   Working must be shown

3. (a) Diagram showing two chlorine atoms with one pair of bonding electrons; must show **all outer** electrons e.g.

   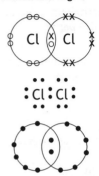

   (b) Tetrahedral/tetrahedron

   (c) Gains an electron (from sodium)

   or

   Indication that the electron arrangement increases by 1

   e.g. Electron arrangement goes from 2.8.7 to 2.8.8, outer electron number goes from 7 to 8

   (d) Low, no

   Both must be correct for 1 mark

   High, no

   Both must be correct for 1 mark

4. (a) (i)

   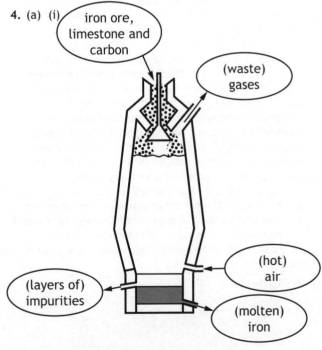

   All must be correct for 2 marks
   2/3 must be correct for 1 mark
   0/1 correct for 0 marks

   (ii) Iron would not melt/be molten/liquid or able to flow

   or

   Iron would be solid

   or

   Iron needs to be molten/liquid/flowing

(b) $Fe^{2+} \rightarrow Fe^{3+} + e^-$

or

$Fe^{2+} - e \rightarrow Fe^{3+}$

**State symbols are not required;
however, if given they must be correct**

5. (a) (i) 14 (days)

**No units required but no mark is awarded if
wrong unit is given. Wrong units would only
be penalised once in the paper**

(ii) 42 (days)

**No units required but maximum of 1 mark is
awarded if wrong unit is given**

Partial marking:

3 half-lives (1)

or

Correct number of days for an incorrect
number of half-lives (1)

Working must be shown

(b) Beta

or

β

or

$_{-1}^{0}β$

or

$_{-1}^{0}e$

or

$_{-1}^{0}e^-$

6. This is an open-ended question.

1 mark: The student has demonstrated a limited
understanding of the chemistry involved. The candidate
has made some statement(s) which is/are relevant
to the situation, showing that at least a little of the
chemistry within the problem is understood.

2 marks: The student has demonstrated a reasonable
understanding of the chemistry involved. The student
makes some statement(s) which is/are relevant to the
situation, showing that the problem is understood.

3 marks: The maximum available mark would be
awarded to a student who has demonstrated a good
understanding of the chemistry involved. The student
shows a good comprehension of the chemistry of the
situation and has provided a logically correct answer to
the question posed. This type of response might include
a statement of the principles involved, a relationship
or an equation, and the application of these to respond
to the problem. This does not mean the answer has to
be what might be termed an "excellent" answer or a
"complete" one.

7. (a) Carboxyl

or

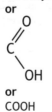

or

COOH

(b) (i) Any acceptable structural formula for butanoic
acid

e.g.

$CH_3CH_2CH_2COOH$

$CH_3(CH_2)_2COOH$

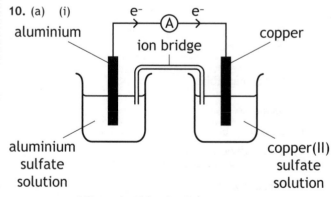

(ii) Butanoic acid or it has bigger/stronger/more
forces (of attraction) (1)
Between molecules or mention of
intermolecular attractions (1)
**If neither of these two points are given, a
maximum of 1 mark can be awarded for:
Butanoic acid or it is bigger/has more carbons
or hydrogens or atoms/longer carbon chain**

8. (a) (i) Glowed brighter/more brightly than zinc

or

Glowed most brightly/very brightly/white light

(ii) Faster/higher/speed up/increase

(b) Magnesium

or

Mg

9. (a) They have the same general formula
and
similar/same chemical properties
Both required for 1 mark

(b) Isomer(s)

(c) (i) Increasing carbon chain length/number of
carbons takes more time (longer, slower)

or

Decreasing carbon chain length/number of
carbons takes less time (faster, quicker)

or

Straight chain takes more time (longer, slower)
than branched chain

or

Branched chain takes less time (faster, quicker)
than straight chain

(ii) Indication that the expected position occurs
anywhere on the horizontal line between
ethane and 2-methylpropane

10. (a) (i)

aluminium

ion bridge

copper

aluminium
sulfate
solution

copper(II)
sulfate
solution

All required for 1 mark

(ii) $3Cu^{2+} + 2Al \rightarrow 3Cu + 2Al^{3+}$
**Accept correct multiples
Zero marks awarded for -electrons shown in
equation, unless clearly scored out
State symbols are not required; however, if
given they must be correct**

(b) 15·79 or 15·8 or 16 (%)

Partial marking:

GFM = 342   (1)

54/342 × 100 (concept mark)   (1)

This step on its own 2 marks

Calculation of final answer using the
correct relationship   (1)

**No units required but maximum of 2 marks
can be awarded if wrong unit is given**

11. (a) pH of solution goes down/decreases/goes
below 7/goes to a value less than 7 from 7 **because**
the H^+ ion/hydrogen ion concentration increases/
goes up or more H^+ than OH^-/$H^+ > OH$

Partial marking:
pH of solution goes down/decreases/goes below 7/
goes to a value less than 7 from 7   (1)
or
H^+ ion/hydrogen ion concentration increases/goes
up/more H^+ than OH^-/$H^+ > OH^-$ (1)

(b) Decreases/goes down/gets lower

12. (a) Select one of these functional groups for 1 mark:

(b) Ester(s)

(c) 21 (g)

Partial marking:
1 mark for **either**
Both GFMs
i.e. 154 and 210
or
Moles of geraniol
i.e. (15·4/154) = 0·1 mol

----

1 concept mark for **either**

$15\cdot4 \times \dfrac{\text{GFM of ester}}{\text{GFM of geraniol}}$

i.e. 15·4 × (210/154)
or
Moles of geraniol × GFM of ester
i.e. 0·1 × 210
**Either of these two steps on their own with all
correct substitutions for 2 marks**
**1 mark for calculated final answer provided the
concept mark has been awarded**
**No units required but a maximum of 2 marks can
be awarded if wrong unit is given**

13. (a) C_nH_{2n-2}
   or
C_nH_{n2-2}
   or
$C_nH_{2(n-1)}$

(b) (i)

(ii) Addition/additional

(c) (i)

(ii) The two bromine atoms are not next to one
another
or
The two bromines are separated by a hydrogen
or
The two bromine branches are not next to one
another

14. (a) (i) Any correct shortened or full structural formula
for hexan-1-ol

(ii) 188 (kJ)

**No units required but no mark is awarded if
wrong unit is given**

(b) 3·9 or 3·91 or 4 (kJ kg^{-1} °C^{-1})
**No units required but a maximum of two marks
can be awarded if wrong unit is given**

Partial marking:
Using the correct concept of
$c = E_h / m\Delta T$
with E_h = 13·3   (1)
0·1 **and** 34   (1)

A further mark can be awarded for the candidate's
calculated answer **only** if the mark for the concept
has been awarded

Alternatively
— 13300 and 0·1 can be used but the final answer
should be 3912 J kg^{-1} °C^{-1} (units must be shown and
correct for 3 marks to be awarded). If no unit, or
the unit given in question is used then 2 marks are
awarded as the mark for the final calculated answer
is not awarded

Or alternatively
— the answer, 3912, can be divided by 1000 to give
the correct answer in kJ kg^{-1} °C^{-1}

Units must be shown and correct for 3 marks to be
awarded. If no unit, or the unit given in question is
used then 2 marks are awarded as the mark for the
final calculated answer is not awarded

NATIONAL 5 CHEMISTRY
2017 SPECIMEN QUESTION PAPER

Section 1

Question	Answer	Max mark
1.	B	1
2.	C	1
3.	A	1
4.	D	1
5.	A	1
6.	D	1
7.	A	1
8.	C	1
9.	C	1
10.	A	1
11.	D	1
12.	A	1
13.	C	1
14.	B	1
15.	A	1
16.	B	1
17.	C	1
18.	D	1
19.	D	1
20.	A	1
21.	B	1
22.	A	1
23.	B	1
24.	C	1
25.	C	1

Section 2

1. (a) 86–90 (seconds)

   Units are not required, but 0 marks should be awarded for the correct answer if incorrect unit is given

   (b) 1.5 cm^3 s^{-1} (3)

   Partial marking:
   1·5 with no unit/incorrect unit   (2)
   $\dfrac{30-0}{20-0}$ or $\dfrac{30}{20}$ or $\dfrac{0-30}{0-20}$ (1)
   Correct unit cm^3 s^{-1} (1)

   (c) Less reactants
   or
   Concentration of reactants decreases
   or
   Reactants are used up
   or
   Less chance of particles colliding
   or
   Equivalent answer

2. (a) Atoms with same atomic number/number of protons/positive particles but different mass number/number of neutrons

   (b) Protons = 35
   Neutrons = 44
   Both required for 1 mark

   (c) Equal amounts/proportions/abundance
   or
   Same number of each
   or
   50:50
   or
   Equivalent answers

   (d) This is an open-ended question.
   1 mark: The candidate has demonstrated a limited understanding of the chemistry involved. The candidate has made a/some statement(s) which is/are relevant to the situation, showing that at least a little of the chemistry within the problem is understood.
   2 marks: The candidate has demonstrated a reasonable understanding of the chemistry involved. The candidate has made a/some statement(s) which is/are relevant to the situation, showing that the problem is understood.
   3 marks: The candidate has demonstrated a good understanding of the chemistry involved. The candidate shows a good comprehension of the chemistry of the situation and has provided a logically correct answer to the question posed. This type of response might include a statement of the principles involved, a relationship or an equation, and the application of these to respond to the problem. This does not mean the answer has to be what might be termed an "excellent" answer or a "complete" one.

3. (a) $Al^{3+}(OH^-)_3$

   (b) (i) 11·2 (pence).

   Unit is not required, but 0 marks can be awarded for the correct answer if incorrect unit is given. Incorrect units would only be penalised once in any paper

   (ii) Named active ingredient with an appropriate reason.

   e.g. magnesium hydroxide
   — cheapest/doesn't fizz

   Aluminium hydroxide — need to take least amount

4. (a) (a higher)

   (b) (i) For appropriate format: scatter graph — i.e. a graph in which points are plotted with their x and y values representing temperature and solubility   (1)

   The axis/axes of the graph has/have suitable scale(s). For the graph paper provided within the question paper, the selection of suitable scales will result in a graph that occupies at least half of the width and half of the height of the graph paper   (1)

   The axes of the graph have suitable labels and units   (1)

   All data points plotted accurately with a line of best fit drawn   (1)

(ii) 10·2 − 10·3 (g/100 cm³)
   or
   A value correctly read from candidate's graph
   (allow ½ box tolerance)

   Units are not required, but 0 marks can be
   awarded for correct answer if incorrect unit is
   given

5. (a) (i) $Li_2CO_3 + 2HCl \rightarrow 2LiCl + CO_2 + H_2O$
   (or correct multiples)

   (ii) LiCl
   or
   Lithium chloride

   (b) (i) 0·44 (g)   (3)

   Units are not required, but a maximum of
   2 marks can be awarded for the correct answer
   if incorrect unit is given

   Partial marking:
   Both *GFMs* 100 and 44   **(1)**

   Correct application of the relationship
   between moles and mass   **(1)**

   This could be shown:

   • by working containing the two expressions

   $$\frac{1}{candidate's\ GFM\ for\ CaCO_3}$$

   and

   $no.\ moles\ CO_2 \times candidate's\ GFM\ CO_2$

   or

   • by working showing correct proportionality
   $$1 \leftrightarrow \frac{candidate\ GFM\ CO_2}{candidate\ GFM\ CaCO_3}$$

   Where the candidate has been awarded
   the mark for the correct application of the
   relationship between moles and mass,
   a further mark can be awarded for correct
   follow through to a final answer   **(1)**

   (ii) Method B   **(1)**

   Gas is lost in method A before starting mass
   taken
   or
   Gas is lost before all acid is added
   or
   No total mass of all reactants at the start of
   experiment
   or
   Equivalent response   **(1)**

6. (a) Flame test (or correct description)
   and
   lilac/purple
   Both required for 1 mark

   (b) Greater than 7
   or
   Any numerical value greater than seven

   (c) Ethene
   Accept correct formula

   (d) Benzoic acid

   (e) Aluminium, silicon and oxygen
   Accept correct formulae

7. (a) Group/family/chemicals/compounds with same
   general formula and same/similar chemical
   properties
   Both parts required for 1 mark

   (b) (i) Diagram showing carbon with four hydrogen
   atoms: each of the four overlap areas must have
   two electrons in or on overlap area (cross, dot,
   petal diagram)
   e.g.

   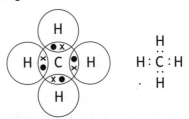

   (ii) Weak bond/attraction   **(1)**
   between molecules   **(1)**

8. (a) Hydroxyl

   (b) Carbon dioxide
   and
   water
   Both required for 1 mark
   Accept correct formulae

   (c) (i) Addition
   or
   Hydration

   (ii)

   H—C—C—C—C—H (with H atoms and O—H, and CH₃ branch)

   or

   H—C—C—C———C—H (with H atoms, O—H and CH₃ branch)

   Accept full or shortened structural formula

   (d) (i)

   H—C—C with =O and O—H

   Accept full or shortened structural formula

   (ii) Carboxylic acid
   Accept alkanoic acid

9. (a) Exothermic

(b) (i) 33·44 (kJ) (3)

Partial marking:
Using $cm\Delta T$ with $c = 4\cdot18$ (1)

To be awarded this concept mark, candidates do not specifically need to write $cm\Delta T$. The concept mark is awarded for using this relationship with three values, one of which must be 4·18

For values 0·2 (kg) and 40 (°C) (1)

A further mark can be awarded for arithmetical follow through to the candidate's answer only if the mark for the $cm\Delta T$ concept has been awarded (1)

Units are not required, but a maximum of 2 marks can be awarded for the correct answer if incorrect unit is given

(ii) Draught insulation
or
Use metal beaker
or
Repeat to get average
or
Any reasonable answer

(c) (i) As the number of carbons increases the energy released increases
or
As the number of carbons decreases the energy released decreases
or
The energy increases as the number of carbons increases
or
The energy decreases as the number of carbons decreases

(ii) 3520 to 3550 (kJ)

Units are not required, 0 marks can be awarded for the correct answer if incorrect unit is given

10. (a)

or

In the formula above, the bond to the methyl group must be correctly aligned with the C atom of the group

(b) Diagram showing delivery tube passing into a test tube which is placed in a water/ice bath

Delivery tube must extend close enough to the neck of the test tube to ensure the vapour can enter the test tube

(c) $C_{10}H_{16}Br_4$

11. (a) Reduction

(b) (i) d.c.

(ii) Negative — (brown solid formed)
Positive — (bubbles of gas)
Both required for one mark

12. (a) 46·67/46·7/47 (3)

Partial marking:
GFM = 60 (1)

$\dfrac{28}{\text{candidate's GFM}} \times 100$ (1)

Calculation of final answer using the relationship

% by mass $= \dfrac{m}{GFM} \times 100$ (1)

(b) (i) Haber (-Bosch)

(ii) Speeds up reaction
or
Less energy/temperature/heat required

(c) Platinum
Accept platinum and rhodium (alloy)

13. (a) (i) Burette

(ii) 16 or 16·0 (cm³)
Units are not required; 0 marks can be awarded for the correct answer if incorrect unit is given

(b) Titration

14. (a) 0·5 (g) (2)
Units are not required but a maximum of 1 mark can be awarded for the correct answer if incorrect unit is given

Partial marking:
1 mark can be awarded for either:
• 2 half lives
or
Mass correctly calculated for an incorrect number of half-lives shown.

(b) Short half-life
or
Would not last long in the body
or
Gamma would go right through body
or
Equivalent response

(c) Beta/β/$_{-1}^{0}e$/$_{-1}^{0}\beta$

The charge on the beta particle does not need to be shown

Do not accept electron without atomic and mass numbers, i.e. e or e-

15. (a) 0·01 (mol) (2)
Units are not required but a maximum of 1 mark can be awarded for the correct answer if incorrect unit is given

Partial marking:
1 mark can be awarded for either
• 143·5 g
or
• Correctly calculated answer for $\dfrac{1\cdot435}{\text{incorrect GFM}}$

(b) 0·5 (mol l^{-1})   (2)

Units are not required but a maximum of 1 mark can be awarded for the correct answer if incorrect unit is given

Partial marking:

1 mark can be awarded for either

- $\dfrac{0·01}{0·02}$

or

- Correctly calculated answer for $\dfrac{0·01}{20}$

If correct relationship is used but volume not converted to litres e.g. 0·01/20 maximum 1 mark

16. This is an open-ended question.

1 mark: The candidate has demonstrated a limited understanding of the chemistry involved. The candidate has made a/some statement(s) which is/are relevant to the situation, showing that at least a little of the chemistry within the problem is understood.

2 marks: The candidate has demonstrated a reasonable understanding of the chemistry involved. The candidate has made a/some statement(s) which is/are relevant to the situation, showing that the problem is understood.

3 marks: The candidate has demonstrated a good understanding of the chemistry involved. The candidate shows a good comprehension of the chemistry of the situation and has provided a logically correct answer to the question posed. This type of response might include a statement of the principles involved, a relationship or an equation, and the application of these to respond to the problem. This does not mean the answer has to be what might be termed an "excellent" answer or a "complete" one.

NATIONAL 5 CHEMISTRY 2018

Section 1

Question	Answer	Mark
1.	A	1
2.	B	1
3.	B	1
4.	D	1
5.	A	1
6.	D	1
7.	C	1
8.	B	1
9.	C	1
10.	D	1
11.	C	1
12.	C	1
13.	A	1
14.	B	1
15.	B	1
16.	C	1
17.	A	1
18.	B	1

19.	A	1
20.	C	1
21.	C	1
22.	D	1
23.	D	1
24.	B	1
25.	D	1

Section 2

1. (a) Carbon dioxide

(b) (i) 0·5 or $\frac{1}{2}$ cm^3 s^{-1} (3)

Partial marking:

0·5 with no unit/incorrect unit   (2)

OR

1 mark awarded for concept of change in volume/change in time.

$\dfrac{77 - 62}{50 - 20}$ or $\dfrac{15}{30}$ or $\dfrac{62 - 77}{20 - 50}$ (1)

Correct unit cm^3 s^{-1} (1)

This mark is independent of the calculated value.

The mark for a final answer can only be awarded if the concept of **change in** volume/**change in** time is correct ie incorrect values **from the table** used (subtractions must be shown and volumes chosen **must** correspond to chosen times).

(ii) One mark is awarded for a graph which shows points plotted rather than bars.   (1)

The axis/axes of the graph has/have suitable scale(s). For the graph paper provided within the question paper, the selection of suitable scales will result in a graph (plotted points) that occupies at least half of the width and half of the height of the graph paper.   (1)

The axes of the graph have suitable labels and units.   (1)

All data points plotted accurately (within a half box tolerance) with either a line of best fit drawn or plots joined.

This mark can only be accessed if linear scales for both axes have been provided.   (1)

(iii) Answer must be correct for the candidate's graph (within a half box tolerance).   (1)

If no graph drawn, 68 ± 1.

Unit is not required; however zero marks are awarded for the correct value with incorrect unit.

(c) Greater number/concentration/moles of hydrogen ions/H$^+$ (1)

OR

more H$^+$ ions.

2. (a) (i) Addition (1)

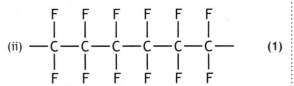

(ii) (1)

(b) Correctly drawn **full** structural formula for ethene. (1)

3. (a) As the percentage of carbon increases the heat content also increases. (1)

OR

As the percentage of carbon decreases the heat content also decreases.

OR

The heat content increases as the percentage of carbon increases.

OR

The heat content decreases as the percentage of carbon decreases.

(b) 46·67/46·7/47 (%) (3)

Partial marking:

$GFM = 120$ (1)

$$\frac{56}{candidate's\ GFM} \times 100$$ (1)

This step on its own is worth 2 marks if the candidate's gfm is 120.

Calculation of final answer using the relationship

$\% \ by \ mass = \frac{m}{GFM} \times 100$ (1)

(Unit is not required, however a maximum of 2 marks can be awarded for the correct value with incorrect unit.)

4. (a) Hydrogen and carbon. (1)

(b) Methylpropane (1)

OR

2-methylpropane

(c) Stronger and intermolecular forces. (1)

(d) 150 to 154°C inclusive. (1)

(Unit is not required; however zero marks are awarded for the correct value with incorrect unit.)

5. (a) Sodium azide, potassium nitrate and silicon dioxide. (1)

(b) Potassium oxide. (1)

(c) SiO_2 (1)

(d) 44 (litres) (1)

(Unit is not required; however zero marks can be awarded for the correct value with incorrect unit.)

6. (a) (i) Two (1)

(ii) 10·8 (2)

11 with working (2)

Partial marking:

$$\frac{(10 \times 20) + (11 \times 80)}{100}$$ (1)

OR

$$\frac{10 \times 20\% + 11 \times 80\%}{100}$$ (1)

(Unit is not required; however a maximum of 1 mark can be awarded for the correct value with incorrect unit.)

(b) $^{14}_{6}C$ (1)

7. (a) (Ionic) Lattice (1)

(b) (i) Ions are free to move. (1)

(ii) Oxidation (1)

(iii) Allows the product(s) to be identified. (1)

OR

To make sure that only one product is produced at each electrode.

OR

To separate the strontium from the chlorine.

8. This is an open-ended question.

1 mark: The student has demonstrated a limited understanding of the chemistry involved. The candidate has made some statement(s) which is/are relevant to the situation, showing that at least a little of the chemistry within the problem is understood.

2 marks: The student has demonstrated a reasonable understanding of the chemistry involved. The student makes some statement(s) which is/are relevant to the situation, showing that the problem is understood.

3 marks: The maximum available mark would be awarded to a student who has demonstrated a good understanding of the chemistry involved. The student shows a good comprehension of the chemistry of the situation and has provided a logically correct answer to the question posed. This type of response might include a statement of the principles involved, a relationship or an equation, and the application of these to respond to the problem. This does not mean the answer has to be what might be termed an "excellent" answer or a "complete" one.

9. (a) (i) (A molecule that contains) carbon to carbon double bond. (1)

OR

C=C

(ii) Bromine (solution)(water)/Br_2 decolourised/discolourised/goes colourless. (1)

(b) 472·8/473 (kJ) (3)

Partial marking:

Using $cm\Delta T$ with $c = 1·97$ (1)

To be awarded this concept mark, candidates do not specifically need to write $cm\Delta T$. The concept mark is awarded for using this relationship with three values, one of which must be 1·97.

For values

1·5 (kg) **and** 160 (°C) (1)

A further mark can be awarded for arithmetical follow through to the candidate's answer **only if the mark for the $cm\Delta T$ concept has been awarded.** (1)

3 marks can be awarded for 472800 or 473000 J. **However the unit (Joules/J) must be given.**

(Unit is not required if answer given in kilojoules, however a maximum of 2 marks can be awarded for the correct value with incorrect unit.)

10. (a) (i) N_2 + $3H_2$ ⇌ $2NH_3$ (1)

(ii) The reaction (it) is reversible (1)

OR

the reaction (it) occurs in both directions

OR

the reaction (it) reaches (or is at) equilibrium.

(b) Diagram showing nitrogen with three hydrogen atoms:

each of the three overlap areas must have two electrons in or on overlap area. (1)

Either the nitrogen or all three hydrogen symbols must be shown.

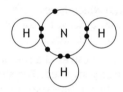

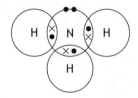

(c) Ostwald (1)

(d) Ammonium nitrate (1)

11. (a) From Mg to Cu through meter/wire. (1)

(b) Completes the circuit/cell (1)

OR

allows ions to flow/move/transfer (between the two beakers)

OR

provide ions to complete the circuit/cell.

(c) $Cu^{2+}(aq) + 2e^- \longrightarrow Cu(s)$ (1)

(d) Insoluble (1)

OR

Solubility less than 1 g l^{-1}

12. (a) (Alpha particles) they (1)

- cannot penetrate the paper
- cannot pass through paper
- are stopped by the paper
- are absorbed by the paper
- can **only** pass through air.

(b) 14·8 (years) (2)

Partial marking:

1 mark can be awarded for **either**:

- 4 half-lives

OR

- number of years correctly calculated for an incorrect number of half-lives (provided the working supports the number of half-lives).

(Unit is not required; however a maximum of 1 mark can be awarded for the correct value with incorrect unit.)

(c) Increases (1)

Stays the same

13. (a) (i) Carboxyl (1)

(ii) 134(g) (1)

(Unit is not required; however 0 marks can be awarded for the correct value with incorrect unit.)

(b) Any correct statement linking acidity to the position of the halogen. (1)

e.g.

The acidity (of the carboxylic acids) decreases going down the group.

OR

As you go (up) from iodine to fluorine the acidity increases.

OR

The one at the top (of the group) has the highest acidity.

OR

The one that has the lowest acidity is at the bottom (of the group).

(c) A correct shortened or full structural formula for (1)

4-methylpentanoic acid

e.g.

$CH_3CH(CH_3)CH_2CH_2COOH$

$HOOCCH_2CH_2CH(CH_3)CH_3$

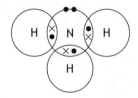

OR

mirror images.

Accept CH_3 for branch in a full structural formula.

14. (a) C (1)

OR

0·24

(b) (i) Pipette (1)

(ii) (2 and 4) They are concordant. (1)

OR

They are within 0·2 (cm^3).

OR

They are within 0·1/0·1 apart.

OR

Titration 1 and 3 or the other two are not concordant or not within 0·2 of each other.

(iii) 0·000161 (mol/mole/moles) (1)

OR

$1·61 \times 10^{-4}$

(Unit is not required; however 0 marks can be awarded for the correct value with incorrect unit.)

15. (a) Diamond(s) anvil cell (1)

(b) (i) 1·9 million (atmospheres) (1)

OR

1 900 000 (atmospheres)

OR

$1·9 \times 10^6$

(Unit is not required; however 0 marks can be awarded for the correct value with incorrect unit.)

(ii) The bulb would not light/would turn off. (1)

(c) $Fe_2O_3 \rightarrow O_2 + Fe_5O_7$ (1)

16. (a) (i) Homologous (series) (1)

(ii) $C_nH_{2n+1}SH/C_nH_{2n+1}HS$ (1)

OR

C_nSH_{2n+2}

OR

$C_nH_{2n+2}S$

OR

$C_nH_{2(n+1)}S$

(b) Sulfur (mon)oxide (1)

Sulfur dioxide

Sulfur trioxide

(c) 960 (g) (3)

Partial marking:

Both *GFM*s 32 and 48 (1)

OR

correct number of moles of methanol (20 moles). (1)

Correct application of the relationship between moles and mass. (1)

This could be shown by:

Method A

moles CH_3SH × candidate's GFM CH_3SH

OR

Method B

by working showing correct proportionality.

$640 \leftrightarrow \dfrac{candidate\ GFM\ CH_3SH}{candidate\ GFM\ CH_3OH} \times 640$

Where the candidate has been awarded the mark for the correct application of the relationship between moles and mass, a further mark can be awarded for correct follow through to a final answer. (1)

OR

Any other valid method accepted.

Unit is note required, however a maximum of two marks can be awarded for the correct value with incorrect unit.

Award zero marks if candidate's working does not use methanol.

17. **1 mark:** The student has demonstrated a limited understanding of the chemistry involved. The candidate has made some statement(s) which is/are relevant to the situation, showing that at least a little of the chemistry within the problem is understood.

2 marks: The student has demonstrated a reasonable understanding of the chemistry involved. The student makes some statement(s) which is/are relevant to the situation, showing that the problem is understood.

3 marks: The maximum available mark would be awarded to a student who has demonstrated a good understanding of the chemistry involved. The student shows a good comprehension of the chemistry of the situation and has provided a logically correct answer to the question posed. This type of response might include a statement of the principles involved, a relationship or an equation, and the application of these to respond to the problem. This does not mean the answer has to be what might be termed an "excellent".

Acknowledgements

Permission has been sought from all relevant copyright holders and Hodder Gibson is grateful for the use of the following:

An extract adapted from 'Hydrogen Storage' from InfoChem Magazine (RSC), Nov 2008 © Royal Society of Chemistry (2017 Section 2 page 6);
An extract adapted from 'Potassium Permanganate' by Simon Cotton from 'Education in Chemistry',
Nov 2009 © Royal Society of Chemistry (2017 SQP Section 2 page 14);
An extract adapted from 'The bizarre world of high pressure chemistry' from 'The Catalyst', Volume 27, Number 1, October 2016, published by the Gatsby Science Enhancement Programme © 2016 STEM Learning Ltd. (2018 Section 2 page 28).